UNIVERSITY OF MARY HARDIN-BAYLOR
3 4637 002 632 064

D0850624

AD HOC NETWORKS
Technologies and Protocols

AD HOC NETWORKS
Technologies and Protocols

Edited by

PRASANT MOHAPATRA
University of California, Davis

SRIKANTH V. KRISHNAMURTHY
University of California, Riverside

Springer

Editors:

Prasant Mohapatra
University of California, Davis
Department of Computer Science
2063 Kemper Hall
Davis, CA 95616
prasant@cs.ucdavis.edu

Srikanth V. Krishnamurthy
University of California, Riverside
Dept. of Computer Science & Engineering
Riverside, CA 92521
krish@cs.ucr.edu

Library of Congress Cataloging-in-Publication Data

Ad HOC networks: technologies and protocols / edited by Prasant Mohapatra, Srikanth Krishnamurthy
p. cm.
Includes bibliographical references and index.
ISBN: 0-387-22689-3 (Printed on acid-free paper.) e-ISBN: 0-387-22690-7
1. Computer networks. I. Mohapatra, Prasant, 1966- II. Krishnamurthy, Srikanth, 1969-

TK5105.5.A29 2004
004.6—dc22 2004051192

Printed in the United States of America.

9 8 7 6 5 4 3 2 SPIN 11515913

springeronline.com

Contents

List of Figures

List of Tables

Contributing Authors

Samir R. Das is an Associate Professor in the Department of Computer Science at the Stony Brook University, New York. His email address is samir@cs.sunysb.edu.

J. J. Garcia-Luna-Aceves is the Baskin Professor of Computer Engineering at the University of California, Santa Cruz, CA. His email address is jj@cse.ucsc.edu.

Mario Gerla is a Professor in the Computer Science Department at the University of California, Los Angeles, CA. His email address is gerla@cs.ucla.edu.

Chao Gui is a doctoral candidate in the Department of Computer Science at the University of California, Davis, CA. His email address is guic@cs.ucdavis.edu.

Robin Kravets is an Assistant Professor in the Department of Computer Science at the University of Illinois in Urbana-Champaign. Her email address is rhk@cs.uiuc.edu.

Prashant Krishnamurthy is an Assistant Professor in the Telecommunications Program at the University of Pittsburgh, PA. His email address is prashant@mail.sis.pitt.edu.

Srikanth Krishnamurthy is an Assistant Professor in the Department of Computer Science and Engineering at the University of California, Riverside, CA. His email address is krish@cs.ucr.edu.

Wenke Lee is an Assistant Professor in the College of Computing at the Georgia Institute of Technology, GA. His email address is wenke@cc.gatech.edu.

Jian Li is a doctoral candidate in the Department of Computer Science at the University of California, Davis, CA. His email address is lijian@cs.ucdavis.edu.

Mahesh K. Marina is a doctoral candidate in the Department of Computer Science at the Stony Brook University, New York. His email address is mahesh@cs.sunysb.edu.

Prasant Mohapatra is a Professor in the Department of Computer Science, University of California, Davis, CA. His email address is prasant@cs.ucdavis.edu.

Seung-Jong Park is a doctoral candidate in the School of Electrical and Computer Engineering at the Georgia Institute of Technology, GA. His email address is sjpark@ece.gatech.edu.

Cigdem Sengul is a graduate student in the Department of Computer Science at the University of Illinois in Urbana-Champaign. Her email address is sengul@uiuc.edu.

Prasun Sinha is an Assistant Professor in the Department of Computer and Information Science at Ohio State University, OH. His email address is prasun@cis.ohio-state.edu.

Raghupathy Sivakumar is an Assistant Professor in the School of Electrical and Computer Engineering at the Georgia Institute of Technology, GA. His email address is siva@ece.gatech.edu.

Karthikeyan Sundaresan is a doctoral candidate in the School of Electrical and Computer Engineering at the Georgia Institute of Technology, GA. His email address is sk@ece.gatech.edu.

Yu Wang is a doctoral candidate in the Department of Computer Engineering at the University of California, Santa Cruz, CA. His email address is ywang@cse.ucsc.edu.

Yongguang Zhang is a Senior Research Scientist at the HRL Laboratories, CA. His email address is ygz@hrl.com.

Preface

Wireless mobile networks and devices are becoming increasingly popular as they provide users access to information and communication anytime and anywhere. Conventional wireless mobile communications are usually supported by a wired fixed infrastructure. A mobile device would use a single-hop wireless radio communication to access a base-station that connects it to the wired infrastructure. In contrast, ad hoc networks does not use any fixed infrastructure. The nodes in a mobile ad hoc network intercommunicate via single-hop and multi-hop paths in a peer-to-peer fashion. Intermediate nodes between a pair of communicating nodes act as routers. Thus the nodes operate both as hosts as well as routers. The nodes in the ad hoc network could be potentially mobile, and so the creation of routing paths is affected by the addition and deletion of nodes. The topology of the network may change randomly, rapidly, and unexpectedly.

Ad hoc networks are useful in many application environments and do not need any infrastructure support. Collaborative computing and communications in smaller areas (building organizations, conferences, etc.) can be set up using ad hoc networking technologies. Communications in battlefields and disaster recovery areas are other examples of application environments. Similarly communications using a network of sensors or using floats over water are other applications. The increasing use of collaborative applications and wireless devices may further add to the need for and the usage of ad hoc networks.

During the last few years, numerous papers and reports have been published on various issues on mobile ad hoc networks. Several tutorials and survey reports have been also published on specific aspects of the mobile ad hoc networks. In fact, conferences and symposiums that are dedicated to ad hoc networking have emerged. However, a "one-stop" resource for overviewing or summarizing the knowledge and progress on ad hoc networking technologies is currently unavailable. Our co-edited book is primarily motivated by these lines of thought.

We have put together a set of interesting chapters that deal with various interesting focal aspects in ad hoc networks. The first chapter is a forerunner for things to come. It primarily motivates the need for ad hoc networks and discusses the evolution of these networks and projects future directions and challenges. The second chapter primarily looks at contention based medium access control in ad hoc networks. Most of the research in ad hoc networks assume the use of either the IEEE MAC protocol or variants thereof and this chapter enuniciates by means of both discussion and analyses the nuances of such MAC protocols. The third chapter provides an in-depth discussion of routing in ad hoc networks. Next, we provide a discussion of multicasting in ad hoc networks, the issues that arise and the technologies that have emerged. We follow with a discussion of transport layer issues and the protocol designs thus far in the fifth chapter. Since ad hoc networks consist of wireless battery operated devices managing energy / power consumption is of paramount importance. The sixth chapter deals exclusively with issues related to power management. Lately, in order to increase the achievable capacity in ad hoc networks there has been a lot of interest in the use of directional antennas and we deliberate various protocols that have emerged for use with such antennas in Chapter seven. Various issues related to the provision of quality of service and mechanisms for dealing with these issues are presented in the eighth chapter. Finally, we have a chapter on security, a vital component that will determine the successful deployment and emergence of ad hoc networks.

PRASANT MOHAPATRA AND SRIKANTH KRISHNAMURTHY

Acknowledgments

We wish to express our heartfelt thanks to all of the authors for helping us with this effort and for participating in this effort. We also wish to thank Chao Gui for helping us with compiling and formating of the chapters. Our thanks to Alex Greene from Kluwer for his patience while we were in the process of completing the book and for his support throughout. Thanks are also due to our families for their support and patience during the entire process.

Chapter 1

AD HOC NETWORKS

Emerging Applications, Design Challenges and Future Opportunities

Mario Gerla
UCLA Computer Science Department
Los Angeles, CA 90095
gerla@cs.ucla.edu

Abstract This book covers the major design issues in ad hoc networks. It will equip the researchers in the field with the essential tools to attack the design of a complex ad hoc system, whether in standalone configurations like the 10,000 node battlefield networks or opportunistically connected to the Internet. The introductory chapter that follows will provide a definition and characterization of ad hoc networks, followed by an overview of the main applications. The design challenges at the various layers of the ad hoc network architecture are then reviewed, with particular emphasis on scalability and mobility. An urban grid scenario that captures the complexities of both standalone and "opportunistically extended" ad hoc designs is introduced. This scenario poses challenges at all the layers of the ad hoc protocol stack. It is thus the ideal framework to illustrate the impact and the key contributions of the various chapters in this book. We conclude with a mention of problems that lie ahead, for further probing by future researchers.

Keywords: Ad Hoc Network, MANET

1.1 Introduction and Definitions

Internet usage has skyrocketed in the last decade, propelled by web and multimedia applications. While the predominant way to access the Internet is still cable or fiber, an increasing number of users now demand **mobile, ubiquitous access** whether they are at work, at home or on the move. For instance, they want to compare prices on the web while shopping at the local department store, access Internet "navigation" aids from their car, read e-mail while riding a bus or hold a project review while at the local coffee shop or in the airport lounge.

The concept of wireless, mobile Internet is not new. When the packet switching technology, the fabric of the Internet, was introduced with the ARPANET in 1969, the Department of Defense immediately understood the potential of a **packet switched radio** technology to interconnect mobile nodes in the battlefield. The DARPA Packet Radio project which began in the early 70's helped establish the notion of **ad hoc wireless networking**. This is a technology that enables untethered, wireless networking in environments where there is no wired or cellular infrastructure (eg, battlefield, disaster recovery, etc); or, if there is an infrastructure, it is not adequate or cost effective.

The term "ad hoc" implies that this network is a network established for a special, often extemporaneous service customized to applications. So, the typical ad hoc network is set up for a limited period of time. The protocols are tuned to the particular application (e.g., send a video stream across the battlefield; find out if a fire has started in the forest; establish a videoconference among 3 teams engaged in a rescue effort). The application may be mobile and the environment may change dynamically. Consequently, the ad hoc protocols must self-configure to adjust to environment, traffic and mission changes. What emerges from these characteristics if the vision of an extremely flexible, malleable and yet robust and formidable network architecture. An architecture that can be used to monitor the habits of birds in their natural habitat, and which, in other circumstances, can be structured to launch deadly attacks onto unsuspecting enemies.

Because of its mobile, non-infrastructure nature, the ad hoc network **poses new design requirements**. The first is **self-configuration** (of addresses and routing) in the face of mobility. At the application level, ad hoc network users typically communicate and collaborate as teams (for example, police, firefighters, medical personnel teams in a search and rescue mission).These applications thus require efficient group communications (**multicasting**) for both data and real time traffic. Moreover, **mobility** stimulates a host of location based services non existent in the wired Internet.

The complexity of mobile ad hoc network designs has challenged generations of researchers since the 70's, Thanks in part to the advances in radio technology, major success have been reported in military as well as civilian applications on this front (eg, battlefield, disaster recovery, homeland defense, etc). At first look, these applications are mutually exclusive with the notion of "infrastructure networks and the Internet" on which most **commercial applications** rely. This is in part the reason why the ad hoc network technology has had a hard time transitioning to commercial scenarios and touching people's everyday lives.

This may soon change, however. An emerging concept that will reverse this trend is the notion of "**opportunistic ad hoc networking**". An opportunistic ad hoc subnet connects to the Internet via "wireless infrastructure" links like 802.11 or 2.5/3G, extending the reach and flexibility of such links. This could be beneficial, for example, in indoor environments to interconnect out of reach devices; in urban environments to establish public wireless meshes which include not only fixed access point but also vehicles and pedestrians, and; in Campus environments to interconnect groups of roaming students and researchers via the Internet. It appear thus that after more than 30 years of independent evolution, ad hoc networking will get a new spin and wired Internet and ad hoc networks will finally come together to produce viable commercial applications.

1.1.1 Wireless Evolution

We begin by studying the evolution of wireless communications systems and networks. The rapid advances of radio technology in the 70's stimulated the development of mobile communications systems that would meet the needs of young professionals on the move. First, there came the need to communicate while on the move, or away from a fixed phone outlet or internet plug. The cellular phone explosion took the original developers by surprise, but it was actually a very predictable phenomenon because telephony is by definition a mobile application. In fact, in our daily life we often use the phone just because we are on the move, for example, we call friends to obtain directions, to coordinate our movements/schedules, etc.

Next, on the heel of the success of cellular telephony came the interest to connect to the Internet from mobile terminals. The traditional Internet applications are less "mobile" than telephony (most of us would prefer to read our e-mail from the convenience of a home than from the road). However, since we are spending an increasing number of hours in cars, trains and planes, we want to fully utilize the travel time with Internet work. New emerging Internet "location based" services (e.g., navigation assistance, store price comparisons, tourist/hotel/parking, etc.) will soon make the wireless connected PDA an indispensable companion. The second wireless wave (mobile Internet access) is supported by wireless LAN technology (predominantly, IEEE 802.11) and by data cellular services (e.g., GPRS, 1xRTT and UMTS); again, a plethora of standards exist also for Data. Both cellular and wireless networking services are supported by an infrastructure and address well established and understood "commodity" needs of the users (e.g., conferencing, e-mail, web access etc).

The third wave in this wireless revolution is the so called "Ad Hoc networking". This type of network was borne with goals very different from mobile telephony and Internet access. The primary goal was to set up communications for specialized, customized, extemporaneous applications in areas where there is no preexisting infrastructure (e.g., jungle explorations, battlefield), or where the infrastructure has failed (e.g., earthquake rescue), or it is not adequate for the current needs (e.g., interconnection of low energy environmental sensors).

With the exception of environment sensor networks (where ad hoc networking is motivated by lack of convenient, low cost infrastructure), most of the other ad hoc applications are "mobile". In fact, they often reflect coordinated mobility patterns (e.g., group motion, swarming, etc.). They involve heterogeneous node types (with different form, energy, transmission range and bandwidth factors); and heterogeneous traffic (voice, data and multimedia). They often pose critical time constraints (because of the multimedia traffic and the emergency nature of the applications).In the following section we review the characteristics of ad hoc networks in more detail.

1.1.2 Ad hoc Networks Characteristics

Mobility: the fact that nodes can be rapidly repositioned and/or move is the raison d'etre of ad hoc networks. Rapid deployment in areas with no infrastructure often implies that the users must explore an area and perhaps form teams/swarms that in turn coordinate among themselves to create a taskforce or a mission. We can have individual random mobility, group mobility, motion along preplanned routes, etc. The mobility model can have major impact on the selection of a routing scheme and can thus influence performance.

Multihopping: a multihop network is a network where the path from source to destination traverses several other nodes. Ad hoc nets often exhibit multiple hops for obstacle negotiation, spectrum reuse, and energy conservation. Battlefield covert operations also favor a sequence of short hops to reduce detection by the enemy.

Self-organization: the ad hoc network must autonomously determine its own configuration parameters including: addressing, routing, clustering, position identification, power control, etc. In some cases, special nodes (e.g., mobile backbone nodes) can coordinate their motion and dynamically distribute in the geographic area to provide coverage of disconnected islands

Energy conservation: most ad hoc nodes (e.g., laptops, PDAs, sensors, etc.) have limited power supply and no capability to generate their own power (e.g., solar panels). Energy efficient protocol design (e.g., MAC, routing, resource discovery, etc) is critical for longevity of the mission.

Scalability: in some applications (e.g., large environmental sensor fabrics, battlefield deployments, urban vehicle grids, etc) the ad hoc network can grow to several thousand nodes. For wireless "infrastructure" networks scalability is simply handled by a hierarchical construction. The limited mobility of infrastructure networks can also be easily handled using Mobile IP or local handoff techniques. In contrast, because of the more extensive mobility and the lack of fixed references, pure ad hoc networks do not tolerate mobile IP or a fixed hierarchy structure. Thus, mobility, jointly with large scale is one of the most critical challenges in ad hoc design.

Security: the challenges of wireless security are well known - ability of the intruders to eavesdrop and jam/spoof the channel. A lot of the work done in general wireless infrastructure networks extends to the ad hoc domain. The ad hoc networks, however, are even more vulnerable to attacks than the infrastructure counterparts. Both active and passive attacks are possible. An active attacker tends to disrupt operations (say, an impostor posing as a legitimate node intercepts control and data packets; reintroduces bogus control packets; damages the routing tables beyond repair; unleashes denial of service attacks, etc.). Due to the complexity of the ad hoc network protocols these active attacks are by far more difficult to detect/fold in ad hoc than infrastructure nets. Passive attacks are unique of ad hoc nets, and can be even more insidious than the active ones. The active attacker is eventually discovered and physically disabled/eliminated. The passive attacker is never discovered by the network. Like a "bug", it is placed in a sensor field or at a street corner. It monitors data and control traffic patterns and thus infers the motion of rescue teams in an urban environment, the redeployment of troops in the field or the evolution of a particular mission. This information is relayed back to the enemy headquarters via special communications channels (eg, satellites or UAVs) with low energy and low probability of detection. Defense from passive attacks require powerful novel encryption techniques coupled with careful network protocol designs.

Unmanned, autonomous vehicles: some of the popular ad hoc network applications require unmanned, robotic components. All nodes in a generic network are of course capable of autonomous networking. When autonomous mobility is also added, there arise some very interesting opportunities for combined networking and motion. For example, Unmanned Airborne Vehicles (UAVs) can cooperate in maintaining a large ground ad hoc network interconnected in spite of physical obstacles, propagation channel irregularities and enemy jamming. Moreover, the UAVs can help meet tight performance constraints "on demand" by proper positioning and antenna beaming.

Connection to the Internet: as earlier discussed, there is merit in extending the infrastructure wireless networks opportunistically with ad hoc appendices. For instance, the reach of a domestic wireless LAN can be extended as needed (to the garage, the car parked in the street, the neighbor's home, etc) with portable routers. These opportunistic extensions are becoming increasingly important and in fact are the most promising evolution pathway to commercial applications. The integration of ad hoc protocols with infrastructure standards is thus becoming a hot issue

1.1.3 Wireless Network Taxonomy

From the above, it is clear that ad hoc nets offer challenges (and opportunities) well beyond the reach of infrastructure networks. So, where do these nets fit in the overall wireless network classification? Most researchers will view ad hoc wireless networks as a special subset of wireless networks. In fact, the ad hoc radio technology and most of the MAC technology will be driven by the advancements in infrastructure wireless networks. The unique design features on ad hoc nets marking a departure from the former are in the network and transport protocol areas (routing, multicast, ad hoc TCP and streaming, etc). Another important family of ad hoc networks, the sensor networks, can in turn be viewed as a subset of ad hoc networks. There are differences, however. At the physical, MAC and network layers, the major innovations and unique features of sensor nets (which set them apart from conventional ad hoc networks) are the miniaturization, the embedding in the application contexts and the compliance with extreme energy constraints. At the application layer, the most unique and novel feature of sensor nets is undoubtedly the integration of transport and in-network processing of the sensed data.

1.2 Ad Hoc Network Applications

Identifying the emerging commercial applications of the ad hoc network technology has always been an elusive proposition at best. Of the three above mentioned wireless technologies - cellular telephony, wireless Internet and ad hoc networks - it is indeed the ad hoc network technology that has been the slowest to materialize, at least in the commercial domain. This is quite surprising since the concept of ad hoc wireless networking was born in the early 70's, just months after the successful deployment of the Arpanet, when the military discover the potential of wireless packet switching. Packet radio systems were deployed much earlier than any cellular and wireless LAN technology. The old folks may still remember that when Bob Metcalf (Xerox Park) came up with the Ethernet in 1976, the word spread that this was one ingenious way to demonstrate "packet radio" technology on a cable!

Why so slow a progress in the development and deployment of commercial ad hoc applications? Main reason is that the original applications scenarios were NOT directed to mass users. In fact, until recently, the driving application was instant deployment in an unfriendly, remote infrastructure-less area. **Battlefield**, Mars explorations, disaster recovery etc. have been an ideal match for those features. Early DARPA packet radio scenarios were consistently featuring dismounted soldiers, tanks and ambulances. A recent extension of the battlefield is the homeland security scenario, where unmanned vehicles (UGVs and UAVs) are rapidly deployed in urban areas hostile to man, say, to establish communications before sending in the agents and medical emergency personnel.

Recently an important new concept has emerged which may help extend ad hoc networking to commercial applications, namely, the concept of **oppor-**

tunistic ad hoc networking. This new trend has been in part prompted by the popularity of wireless telephony and wireless LANs, and the recognition that these techniques have their limits. The ad hoc network is used "opportunistically" to extend a home or Campus network to areas not easily reached by the above; or, to tie together Internet islands when the infrastructure is cut into pieces - by natural forces or terrorists for examples).

Another important area that has propelled the ad hoc concept is **sensor nets**. Sensor nets combine transport and processing and amplify the need for low energy operation, low form factor and low cost - so, these are specialized ad hoc solutions. Nevertheless, they represent a very important growing market.

In the sequel we elaborate on two applications, the **battlefield** and the **the urban and Campus grid**.

1.2.1 The Battlefield

In future battlefield operations, autonomous agents such as Unmanned Ground Vehicles (UGVs) and Unmanned Airborne Vehicles (UAVs) will be projected to the forefront for intelligence, surveillance, strike, enemy antiaircraft suppression, damage assessment, search and rescue and other tactical operations. The agents will be organized in clusters (teams) of small unmanned ground, sea and airborne vehicles in order to launch complex missions that comprise several such teams. Examples of missions include: coordinated aerial sweep of vast urban/suburban areas to track suspects; search and rescue operations in unfriendly areas (e.g., chemical spills, fires, etc), exploration of remote planets, reconnaissance of enemy field in the battle theater, etc. In those applications, many different types of Unmanned Vehicles (UVs) will be required, each equipped with different sensor, video reconnaissance, communications support and weapon functions. A UV team may be homogeneous (e.g., all sensor UVs) or heterogeneous (i.e., weapon carrying UVs intermixed with reconnaissance UVs etc). Moreover, some teams may be airborne, other ground, sea and possibly underwater based. As the mission evolves, teams are reconfigured and individual UVs move from one team to another to meet dynamically changing requirements. In fact, missions will be empowered with an increasing degree of autonomy. For instance, multiple UV teams collectively will determine the best way to sweep a mine field, or the best strategy to eliminate an air defense system. The successful, distributed management of the mission will require efficient, reliable, low latency communications within members of each team, across teams and to a manned command post. In particular, future naval missions at sea or shore will require effective and intelligent utilization of real-time information and sensory data to assess unpredictable situations, identify and track hostile targets, make rapid decisions, and robustly influence, control, and monitor various aspects of the theater of operation. Littoral missions are expected to be highly dynamic and unpredictable. Communication interruption and delay are likely, and active deception and jamming are anticipated.

The Office of Naval Research (ONR) is currently investigating efficient system solutions to address the above problems. ONR envisions unmanned systems of Intelligent, Autonomous Networked Agents (AINS) to have a profound influence on future naval operations allowing continuous forward yet unobtrusive presence and the capability to influence events ashore as required. Unmanned vehicles have proven to be valuable in gathering tactical intelligence by surveillance of the battlefield. For example, UAVs such as Predator and Global Hawk are rapidly becoming integral part of military surveillance and reconnaissance operations. The goal is to expand the UAV operational capabilities to include not only surveillance and reconnaissance, but also strike and support mission (e.g., command, control, and communications in the battle space). This new class of autonomous vehicles is foreseen as being intelligent, collaborative, recoverable, and highly maneuverable in support of future naval operations.

In a complex and large scale system of unmanned agents, such as designed to handle a battlefield scenario, a terrorist attack situation or a nuclear disaster, there may be several missions going on simultaneously in the same theater. A particular mission is "embedded" in a much larger "system of systems". In such a large scale scenario the wireless, ad hoc communications among the teams are supported by a global network infrastructure (the "Internet in the sky"). The global network is provisioned independently of the missions themselves, but it can opportunistically use several of the missions' assets (ground, sea or airborne) to maintain multihop connectivity

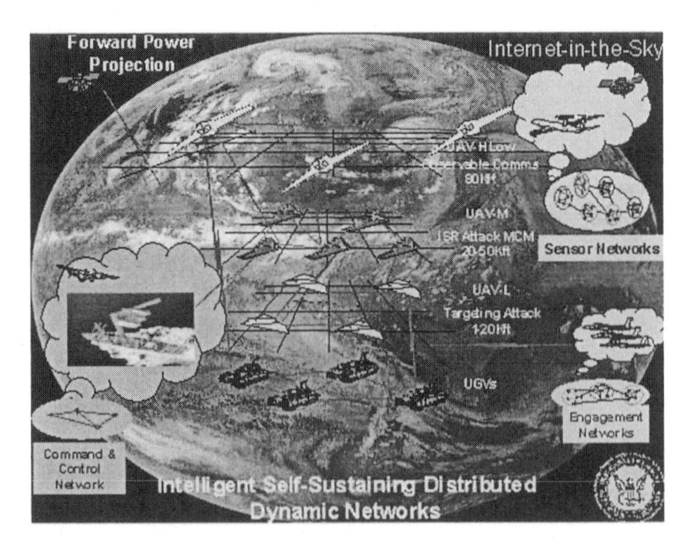

Figure 1.1. Internet in the sky architecture designed as part of the ONR supported Minuteman project at UCLA.

The development of the Internet in the Sky hinges on three essential technologies:

1 Robust wireless connectivity and dynamic networking of autonomous unmanned vehicles and agents.

2 Intelligent agents including: mobile codes, distributed databases and libraries, robots, intelligent routers, control protocols, dynamic services, semantic brokers, message-passing entities.

3 Decentralized hierarchical agent-based organization.

As Figure 1.1 illustrates, the autonomous agents have varying domains of responsibility at different levels of the hierarchy. For example, clusters of UAVs operating at low altitude (1K-20K feet) may perform combat missions with a focus on target identification, combat support, and close-in weapons deployment. Mid-altitude clusters (20-50K feet) could execute knowledge acquisition, for example, surveillance and reconnaissance missions such as detecting objects of interest, performing sensor fusion/integration, coordinating low-altitude vehicle deployments, and medium-range weapons support. The high altitude cluster(s) (50K-80K feet) provides the connectivity. At this layer, the cluster(s) has a wide view of the theater and would be positioned to provide maximum communications coverage and will support high-bandwidth robust connectivity to command and control elements located over-the-horizon from the littoral/targeted areas.

We use this example to focus on mission oriented communications and more precisely on a particular aspect of it, **team multicast**. In team multicast the multicast group does not consist of individual members, rather, of teams. For example, a team may be a special task force that is part of a search and rescue mission. The message then must be broadcast to the various teams that are part of the multicast group, and, to all UVs within each team. For example, a weapon carrying airborne UV may broadcast an image of the target (say, a poison gas plant) to the reconnaissance and sensor teams in front of the formation, in order to get a more precise fix on the location of the target. The sensor UV team(s) that has acquired such information will return the precise location. As another example, suppose N teams with chemical sensors are assessing the "plume" of a chemical spill from different directions. It will be important for each team to broadcast its findings step by step to the other teams using team multicast. In general, team multicast will be common place in ad hoc networks designed to support collective tasks, such as occur in emergency recovery or in the battlefield.

1.2.2 The Urban and Campus Grids: a case for opportunistic ad hoc networking

In this section we describe two sample applications that illustrate the research challenges and the potential power of ad hoc as opportunistic extension of the wireless infrastructure.

Two emerging wireless network scenarios that will soon become part of our daily routines are **vehicle communications** in an urban environment, and **Campus nomadic networking**. These environments are ripe for benefiting from the technologies discussed in this report. Today, cars connect to the cellular system, mostly for telephony services. The emerging technologies however, will soon stimulate an explosion of new applications. **Within the car**, short range wireless communications (e.g., **PAN technology**) will be used for monitoring and controlling the vehicle's mechanical components as well as for connecting the driver's headset to the cellular phone. Another set of innovative applications stems from communications **with other cars** on the road. The potential applications include road safety messages, coordinated navigation, network video games, and other peer-to-peer interactions. These network needs can be efficiently supported by an "opportunistic" **multihop** wireless network among cars which spans the urban road grid and which extends to intercity highways. This ad hoc network can alleviate the overload of the fixed wireless infrastructures (3G and hotspot networks). It can also offer an emergency backup in case of massive fixed infrastructure failure (e.g., terrorist attack, act of war, natural or industrial disaster, etc). The coupling of car multihop network, on-board PAN and cellular wireless infrastructure represents a good example of **hybrid wireless network** aimed at cost savings, performance improvements and enhanced resilience to failures. An example of such network is illustrated in Figure 1.2.

In the above application the vehicle is a communications hub where the extensive resources of the fixed radio infrastructure and the highly mobile ad hoc radio capabilities meet to provide the necessary services. New networking and radio technologies are needed when operations occur in the "extreme" conditions, namely, extreme mobility (radio and networking), strict delay attributes for safety applications (networking and radio), flexible resource management and reliability (adaptive networks), and extreme throughput (radios). Extremely flexible radio implementations are needed to realize this goal. Moreover, cross layer adaptation is necessary to explore the tradeoffs between transmission rate, reliability, and error control in these environments and to allow the network to gradually adapt as the channel and the application behaviors are better appraised through measurements.

Another interesting scenario is the Campus, where the term "Campus" here takes the more general meaning of a place where people congregate for various

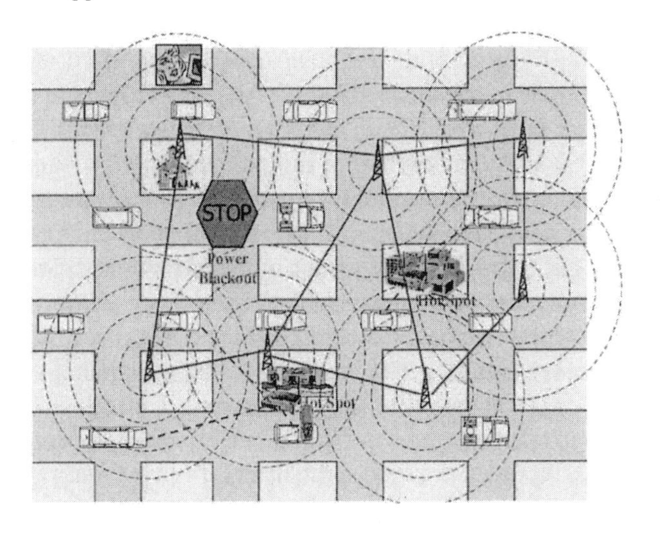

Figure 1.2. An example opportunistic ad hoc network.

cultural and social (possibly group) activities, thus including Amusement Park, Industrial Campus, Shopping Mall, etc. On a typical Campus today wireless LAN access points in shops, hallways, street crossings, etc., enable nomadic access to the Internet from various portable devices (e.g., laptops, notebooks, PDAs, etc.). However, not all areas of a Campus or Mall are covered by department/shop wireless LANs. Thus, other wireless media (e.g., GPRS, 1xRTT, 3G) may become useful to fill the gaps. There is a clear opportunity for multiple interfaces or agile radios that can automatically connect to the best available service. The Campus will also be ideal environment where group networking will emerge. For example, on a University Campus students will form small workgroups to exchange files and to share presentations, results, etc. In an Amusement Park groups of young visitors will interconnect to play network games, etc. Their parents will network to exchange photo shots and video clips. To satisfy this type of close range networking applications, Personal Area Networks such as Bluetooth and IEEE 802.15 may be brought into the picture. Finally, "opportunistic" ad hoc networking will become a cost-effective alternative to extend the coverage of access points. Again, as already observed in the vehicular network example, the above "extensions" of the basic infrastructure network model require exactly the technologies recommended in this report, namely: multimode radios, cross layer interaction (to select the best radio interface) and some form of hybrid networking.

These are just simple examples of networked, mobile applications drawn from our everyday lives. There is a wealth of more sophisticated and demanding applications (for example, in the areas of pervasive computing, sensor networks, battlefield, civilian preparedness, disaster recovery, etc) that will soon be enabled and spun off by the new radio and network technologies.

1.3 Design Challenges

As mentioned earlier, ad hoc networks pose a host of new research problems with respect to conventional wireless infrastructure networks. This book in fact addresses these challenges and each chapter is focused on a particular design issue at one of the layers of the protocol stack. We will provide a review of the chapters shortly. First, we wish to report on some design challenges that cut across the layers and should be kept in mind while reading about specific layer solutions in the other chapters. These are: cross layer interaction; mobility, and; scalability.

1.3.1 Cross Layer Interaction

Cross Layer Interaction/Optimization is a loaded word today, with many different meanings. In ad hoc networks it is however a very appropriate way to refer the fact that it is virtually impossible to design a "universal" protocol (routing, MAC, multicast, transport, etc) and expect that it will function correctly and efficiently in all situations. In fact, pre-defined protocol layers a' la Internet work reasonably well in wired nets (e.g., routing, addressing, DNS etc work for large and small.). For example, the physical and MAC layers of the wired E-net are the uncontested reference for of all Internet designs. In contrast, in the wireless LAN (the closest relative of the E-net), there is convergence not to one, but to a family of standards, from 802.16 to 15 to 11, each standard addressing different environments etc. Even within the 802.11 family a broad range of versions have been defined, to address different needs.

In ad hoc network design the importance of tuning the network protocols to the radios and the applications to the network protocols is even more critical, given the extreme range of variability of the systems parameters. Clearly, the routing scheme that works best for network of a dozen students roaming the Campus may not be suitable for the urban grid with thousand of cars or the battlefield with an extreme range of node speeds and capabilities. Even more important is the concept that in these cases the MAC, routing and applications must be jointly designed. Moreover, as some parameters (eg, radio propagation, hostile interference, traffic demands, etc) may dynamically change, the protocols must be adaptively tuned. Proper tuning requires exchange of information across layers. For example in a MIMO (Multi Input, Multi Output) radio system the antenna and MAC parameters and possibly routes are dynamically reconfigured based on the state of the channel, which is learned from periodic channel measurements . Thus, interaction between radio channel and protocols is mandatory to achieve an efficient operating point. Video adaptation is another example of cross layer interaction: the video rate stipulated at session initialization cannot be maintained if channel conditions deteriorate. The proper rate adjustment requires careful interplay of end to end probing (eg, RTCP) as well measurements from channel and routing.

1.3.2 Mobility and Scaling

Mobility and reconfiguration is what uniquely distinguished ad hoc networks from other networks. Thus, being able to cope with nodes in motion is an essential requirement. Large scale is also common in ad hoc networks, as battlefield and emergency recovery operations often involve thousands of nodes. The two aspects - mobility and scale - are actually intertwined: anybody can find a workable ad hoc routing solution, say, for 10 nodes, no matter how fast they move; and anybody can find a workable (albeit inefficient) solution (for routing, addressing, service discovery etc) for a completely static ad hoc network with 10,000 of nodes, say (just consider the Internet)! The problems arise when the 10,000 nodes move at various speeds, in various directions over a heterogeneous terrain. In this case, a fixed routing hierarchy such as in the Internet does not work. That is when you have to take out the "big guns" to handle the problem.

Mobility is often viewed as the #1 enemy of the wireless ad hoc network designer. However, mobility, if properly characterized, modeled, predicted and taken into account, can be of tremendous help in the design of scaleable protocols. In the sequel we offer a few examples where mobility actually helps.

1.3.2.1 An example: Team Communications among Airborne Agents using LANMAR. LANMAR is a scalable routing protocol for large, mobile, "flat" ad hoc wireless networks. It has been implemented in the Minuteman network under ONR support [1]. LANMAR assumes that the network is grouped into logical subnets in which the members have a commonality of interests and are likely to move as a "group" (e.g., a team of co-workers at a convention; or tanks in a battalion, or UAVs in an unmanned scouting mission). The logical groups are efficiently reflected in the addressing scheme. We assume that a two level, IP like MANET (Mobile Ad hoc NET) address is used consisting of a group ID (or subnet ID) and a host ID, i.e. <Group ID, Host ID>. The group ID tells us which nodes are part of the same group. Group assocoation may change from time to time as a node is reassigned to a different group (e.g. task force in a military scenario). The Host ID is fixed and typically corresponds to the hardwired device address. Such MANET address uniquely identifies the role (and position) of each node in the network. Similar to an IP network, the packet is routed to the group first, and then to the Host within the group. The challenge is to "find" the group in a large, mobile network.

LANMAR uses the notion of landmarks to keep track of such logical groups. Each logical group has one node serving as "landmark". The landmark advertises the route to itself by propagating a Distance Vector, e.g. DSDV (Destination Sequences Distance Vector) [3]. Further, the LANMAR routing scheme is always combined with a local routing algorithm, e.g. Fisheye State Routing (FSR) [2]. FSR is a link state routing algorithm with limited "scope" feature for local, low overhead operation. Namely, FSR knows the routes to all nodes within a predefined Fisheye scope (e.g., 3 hops) from the source. For nodes outside of the Fisheye scope, the landmark distance vector must be inspected for directions. As a result, each node has detailed topology information about nodes within its Fisheye scope and knows distance and routing vector (i.e., direction) to all landmarks. An example of LANMAR routing implementation is shown in Figure 1.3.

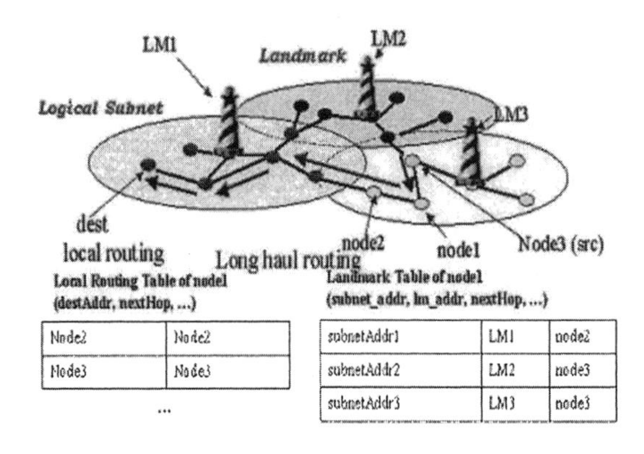

Figure 1.3. An example of LANMAR implementation.

When a node needs to relay a packet to a destination that is within its Fisheye scope, it obtains accurate routing information from the Fisheye Routing Tables. The packet will be forwarded directly. Otherwise, the packet will be routed towards the landmark corresponding to the destination logical subnet, which is read from the logical address field in the MANET address. Thus, when the packet arrives within the scope of the destination, it may be routed to it directly without ever going through the landmark. In summary, the hierarchical LANMAR setup does the scalability trick - it reduces routing table size and route update overhead making the scheme practical for a network with practically unlimited number of nodes (as long as nodes move in groups of increasing size).The latter assumption is actually well validated in ad hoc networks associated with large scale, cooperative operations (eg, battlefield). If nodes are moving randomly and in a non coordinated fashion (like perhaps the customers in a shopping mall) other techniques can be used to achieve scalability in a random motion scenario. Along these lines, recently proposed routing and resource discovery schemes such as "last encounter routing", and "epidemic dissemination" exploit the fact that, with random motion, the destination that I want to reach "has been seen" some time ago by some nodes that now have moved close to me. This is a perfect example of symbiosis of mechanical information transport and electronic information relay. It allows me to find the destination through a "motion assisted" search which eliminates the need for a costly (and definitely non scalable) full search.

1.4 Evaluating Ad Hoc Network Protocols - the Case for a Testbed

Analysis, simulation, hybrid simulation and testbed measurements are well known techniques for evaluating ad hoc network protocols. At a time when ad hoc network "standards" are being proposed in the MANET (Mobile Ad Hoc Networks) working group of the IETF, it is clearly important to have a set of reliable performance evaluation and measurement tools to compare various proposals in a consistent environment that can be calibrated and replicated. This is where the notion of "national" ad hoc network test-bed comes in the picture. In this section we review the mission and goals of one such testbed, the WHYNET NSF Testbed recently established in southern California with the participation of various academic and industrial Campuses.

WHYNET is a wireless networking testbed that can be used to evaluate the impact of emerging technologies that are going to shape the nature of wireless, mobile communications in the next decade. The eventual impact of this research testbed will be to redefine how specific innovations in wireless communication technologies are evaluated in terms of their potential to improve application-level performance as well as how alternative approaches are compared with each other.

WHYNET differs from existing testbeds both in its *scope* and *approach.* Its primary objective is to provide researchers at every layer of the protocol stack, from physical devices to transport protocols, a testbed to evaluate the impact of their technology on application level performance, using scalable and realistic operational scenarios. To achieve this objective, WHYNET will use a geographically-distributed, hybrid networking testbed that combines the realism of physical testing with the scalability of multi-mode simulations.

The primary deliverable from WHYNET will be a set of tools and methodologies encapsulated in a well-defined evaluation framework, a set of studies that demonstrate its suitability for evaluation of emerging network technologies, and a repository of networking scenarios, measurements, and models. The design and development of the testbed will require coordinated efforts of a multi-disciplinary, multi-institution team of researchers from academia, government, and industry. This effort will substantially leverage existing networking research funded by NSF, ONR, ARO, DARPA, and corporate sponsors that include HP, Intel, Ericsson, Nokia, and Microsoft.

A central component of WHYNET will be its incorporation of geographically distributed physical testbeds. This will allow researchers to experiment locally with physical prototypes, while providing a cost effective method to support diverse operational environments in the testbed. The geographically-distributed physical testbed will also be integrated into a scalable, multi-tool simulation framework, which will allow investigators to evaluate the scalability properties of innovative networking technologies. When fully deployed, WHYNET will include a physical 3G CDMA testbed, a multiplicity of radio platforms that include narrowband, broadband, and software defined radios, a set of small to medium physical MANET testbeds incorporating novel radio devices, a collection of measurements and models for a diverse set of antenna and channel conditions, and a large set of reusable protocol models and application scenarios. In addition, WHYNET will be used to perform a set of studies that are expected to include the following:

- Perceptual evaluation of networking protocols
- CLI (Cross Layer Interaction) aware wireless networking
- Comparative evaluation of new radio devices
- Policy based routing with QoS assurance
- Protocols and middleware services for mesh networking
- Sensor networks
- Energy-aware networks
- Security in scalable ad hoc networks
- Adaptive transport protocols

Although the primary purpose of these studies is to evaluate novel networking technologies, they will also be used to demonstrate the unique contributions of testbeds such as WHYNET in the design and evaluation of next generation networking technologies. For instance, the studies on protocols for mesh networking will demonstrate WHYNET capabilities of supporting smooth transition from system design to deployment. Protocol prototypes can communicate with simulated low layers for repeatable results, or obtain varying rate real multimedia application traffic for perceptual evaluation. Once the physical hardware devices are ready for testing, a portion of target network system can be configured with real devices while the rest of the network can still reside in the simulated hardware domain.

1.5 Overview of the Chapters in this Book

In this section we review the chapters of the book, commenting on their specific contributions. In order to relate the contributions to the "big picture", we plan to illustrate their impact on a representative application. In sect 2.6 we depicted the urban grid scenario which provided an excellent example of "opportunistic" ad hoc network. In fact, the urban grid network poses formidable protocol design challenges, from the MAC layer all the way to applications. This book will certainly offer invaluable help to anyone who plans to engage in grid network design, and more generally, in ad hoc network research. To illustrate the relevance of the concepts presented in these chapters, for each chapter that is being reviewed, we will pose the question: How can this suite of protocols help in the design of an urban vehicle grid? The proposed protocols may not answer all the questions. The deal then is to discuss the additional requirements in the Future Research section.

Chapter 2: Collision Avoidance Protocols

This chapter provides an excellent overview of the CSMA/CA protocol along with elegant analytic methods to evaluate the efficiency of the protocol in various scenarios and for various parameters. An additional bonus of this chapter is the discussion of fairness of the MAC layer under UDP (say, for video streaming applications) as well as under TCP. Considering our strawman urban grid application, accurate MAC layer modeling will be critical in the design of the emerging vehicular MAC standards. In particular, it will be important that whatever MAC standard is chosen, it perform well under TCP and streaming. The material in this chapter will assist in that choice.

Chapter 3: Routing in Ad Hoc Networks

This chapter describes various routing protocols that have been proposed for ad hoc networks. Proactive (DSDV, OLSR, TBRPF), and reactive routing protocols (DSR, AODV) and hybrid protocols (ZRP) are evaluated. Particularly interesting is the discussion of geo-routing protocols and more generally, location assisted routing protocols (GPSR, LAR, DREAM). In the urban grid environment cars and pedestrians know their coordinates, thus they can rely on the geographical routing assistance. Hybrid routing may also be considered, in order not to get bogged down too often by the numerous obstacles. This chapter provides the right information to tackle the routing design and evaluation problem.

Chapter 4: Multicasting in Ad Hoc Networks

Multicast (both reliable data multicast and multimedia streaming) is a critical service in MANETs where data and video must be broadcast to all users/teams participating in the same mission (e.g., search and rescue operation). This chapter does a thorough survey of the literature. It also brings up the challenge of node mobility and network dynamic. The most popular multicast protocols - MAODV, ODMRP - are first reviewed. Then, more specialized protocols are introduced: MCEDAR (using the concepts of clustering and backbone), AMRoute (relying on the overlay multicast concept), Geocast, Gossip (based on random re-broadcast). Additional requirements may be placed on top of basic multicast, for example: reliability, QoS, security. Considering our urban grid model, it is easy to visualize the case where a squad of patrol cars, distributed all over town, is engaged in a sweep operation, say looking for a suspect. Any of the above schemes should be carefully evaluated for the urban grid implementation. Naturally, if the multicast group member locations (either GPS or urban grid coordinates) are known, the geocast option becomes very attractive. If the operation is a covert operation, secure multicast is needed to encrypt the contents and also to maintain motion secrecy.

Chapter 5: Transport Layer Protocols in Ad Hoc Networks

TCP accounts for 90% of the traffic in the internet. This trend will be maintained in the a hoc network (unless one goes about a radical change of all the applications). TCP is well known to degrade in mobile ad hoc networks. This chapter analyses the causes of performance degradation. The most obvious indication that something is going wrong is packet loss. However, the loss may be due to congestion - in which case the TCP should slow down. Or it may be caused by random errors, jamming, route breakup induced by motion. In the latter cases, TCP must not slow down the flow, else matters get worse! One well known problem is the inability to discriminate between congestion and random loss. ELFN (Explicit Link Failure Notification) is a network feedback technique that can be used to notify the TCP source of link failure (i.e., no congestion!). The source then refreshes the path while freezing TCP. ATRA is a more elaborate method that tries to minimize the effect of route failure by "predicting" and averting it using aggressive route recomputations. ATP requires a complete redesign of the TCP protocol (using ATM style virtual circuit rate control methods) to take advantage of selective feedback from specific nodes along the path. Not clear how ATP will survive high mobility. In considering the application of these options to the urban grid, one important requirement is the compatibility of ad hoc TCP with the Internet TCP (since traffic may originate or be directed to hosts in the Internet). This seems to rule out ATP immediately since in ATP both source and destination TCP stacks are modified. The remaining schemes are feasible and should be carefully evaluated.

Chapter 6: Energy conservation

In ad hoc networks consisting of moving nodes (e.g. vehicles), energy conservation is generally not a critical issue. However, it clearly becomes a concern in sensor networks or in ad hoc networks where the time to discharge a "powered-on" node is less than the time between battery recharging opportunities. This chapter provides an excellent survey of the various techniques to conserve power, namely: power/topology control, energy routing, coordinated sleep and power save management. If we go back to our urban grid example, we note that cars have a practically unlimited reserve of energy. However, pedestrians do not, especially if they use 802.11 in their PDAs. If the PDA has multiple interfaces, say 802.11, ZigBee, cellular and Bluetooth, all the latter options are more attractive as "always - on" options instead of 802.11. In fact, radio interface selection could be yet another energy conservation strategy to add to the above list. Another important component in the urban grid is the environment sensor fabric. These sensors must interact with pedestrians and cars (for example, a sensor field comes alive if a police car approaches). Thus, sensors (and pedestrians) must be scheduled in such a way that their interaction is most effective for a given recharge cycle. The schemes described in this chapter are an excellent start for the investigation of suitable sensor/pedestrian energy strategies.

Chapter 7: Use of Smart Antennas in Ad Hoc Networks

Directive antennas are used for at least three reasons: extending range, folding jamming attacks and reducing the probability of detection. Smart antennas add another feature - the ability to transmit simultaneously on multiple beams. This chapter gives a brief overview of directional antennas. It then provides an exhaustive survey of the interaction between antenna beamforming, MAC protocols and routing protocols. It is in fact clear that, to take advantage of antenna directionality, MAC and routing protocol changes are required. Are smart antennas going to have an impact on our urban grid network strategy. Absolutely! One can take advantage of the extended range of directional antennas to establish backbone links along the major boulevards, say. Also, if UAVs are used to assist in urban disaster recovery, directional antennas will do very well for ground to air and air to air links. One important issue indirectly addressed by this chapter is the coexistence of different MAC and routing protocols in the same network, since only part of the nodes will be capable of antenna beaming. In all, this chapter is an excellent start for an investigation of mixed antenna strategies in complex environments such as urban grids and battlefields.

Chapter 8: QoS Issues in ad hoc networks

QoS support is critical in ad hoc networks since such networks either operate as "opportunistic" extensions of the internet and thus carry Internet multimedia traffic (VoIP, videocast, videoconference, etc); or, they operate in emergency mode, and have even more stringent QoS requirements (delay, latency, jitter, packet loss, etc)! This chapter does an excellent job in explaining the difference between QoS guarantees in wired and in wireless ad hoc networks. It begins by reviewing the methods for improving the performance of the 802.11 physical layer (ARF, RBAR, OAR) and its impact on QoS. It then moves to the MAC layer and shows how the 802.11b and 802.11e mechanisms (e.g., PCF schedule, IFS, etc) can be manipulated to achieve DiffServ type PHB (Per Hop Behavior). This is followed by a discussion of QoS routing which allows the source to enforce Call Acceptance Control and/or service negotiation. INSIGNIA signaling could be used for such negotiation. All this is body of information is very relevant to our urban grid network. Suppose you want to watch a soccer game in your car. Should you receive over the ad hoc car-net for free, or from UMTS and pay a connection fee. The ad hoc network QoS mechanisms will tell your "intelligent" mobile middleware which options are available, and for how long (if you buy the predictive location based routing protocol described in this chapter!). After you decide to use the ad hoc network (to save $$$!!), the MAC and physical layer parameters will be set to match your DiffServ DSCPs. Routing will abide to its promise and find the route that fits your request.

Chapter 9: Security in mobile Ad Hoc Networks

Ad hoc networks are much more vulnerable to security attacks than conventional wired networks. The reasons: open wireless medium; capture of unattended roaming nodes and impersonation; decentralized coordination protocols vulnerable to attack (e.g., contention based MAC); lack of centralized certificate authority for key exchange; use of cache proxies that can be easily hit by DDoS attacks, etc. This chapter reviews the various types of possible attacks and discusses prevention measures. It introduces a MANET architecture with Intrusion Detection System (IDS) agents located at monitoring nodes, and dwells on the possible IDS agent cooperation strategies. This IDS technique is then applied to detect of an attack to on demand routing (DSR or AODV) by "anomaly" detection. In the context of our strawman urban grid scenario, the protection from attacks is critical. MAC and routing attacks by a terrorist group, for example, if successful, could impair the communications among the police agents that try to apprehend them. Naturally, there are also "passive" attacks we must protect from, for example position and motion privacy attacks. This is an extremely important area, for which this chapter represents an excellent introduction.

1.6 Conclusions

This book offers a solid background in ad hoc network protocols and technologies from which students and researchers can spring forward and attack future challenges in the field. Among these future challenges for further probing we mention:

1. **Wired and wireless interconnection:** the 4G architecture will consist of the interconnection of various wireless technologies with each other and with the wired infrastructure. An important issue will be to interconnect ad hoc network islands with the wired network. For example, the interconnection of ad hoc Campus networks via the Internet in such a way that the ad hoc network users are unaware of the wired network. Critical issues will be scalability, transparency and smooth handoff.

2. **Backbone network:** scalability is the major limitation to large scale deployment of ad hoc networks. One way to solve the problem is to use the existing infrastructure (eg, Internet, satellites, etc). If there is no infrastructure, an important research direction is the use of mobile backbone nodes.

3. **Sensor integration with the ad hoc network:** today, sensor networks are developed and deployed with unique protocols and radio technologies suitable for low energy operations and for the unique processing needs of sensor nets - low energy, in-network processing, propagation of alarms to collection centers. The information collected and processed by the sensor fabric must often be relayed remotely to decision centers via an ad hoc network. For example, in a heavily instrumented battlefield UAVs and UGVs may be dispatched to extract information from the sensor fields and make it available in the ad hoc network. This will require careful coordination of sensor and network protocols. For example, content based addressing instead of IP addressing will be the norm.

4. **Exploiting mobility:** node mobility if attacked in brute force mode can be a serious obstacle to scalability, security and QoS support. However, mobility can be exploited to make our job easier. The advantages of accounting for group mobility were already exposed in sect 3.2.1 of this chapter (LANMAR protocol). Other important benefits are in epidemic diffusion of indices and "last encounter" routing. Motion prediction can also assist in making georouting more efficient. Similarly, the presence of high performance access points (eg, infostations or backbone nodes) on a node's trajectory may encourage to delay a data transfer instead of transmitting the data immediately to low power neighbors.

5 **Motion privacy:** security in wireless networks today mainly addresses the protection of content and the defense from active attacks (internal or external). An insidious passive attack that has mostly passed unnoticed is the location and motion privacy attack. A mobile node may not wish others to track its location or motion. Yet, the mere use of the most popular routing protocols (e.g., AODV, OLSR, DSR etc) can easily give away all the position and motion information to a "passive" intruder which (being passive) will never be caught! This is particularly critical for covert operations in the battlefield or in urban emergencies. The key to protection is to embed security in our MANET protocols directly.

The above is just a small sample of the problems that lie ahead and await you after you muster the content of this book. Enjoy the reading and be prepared for ever greater challenges.

References

[1] M. Gerla, X. Hong, and G. Pei. Landmark routing for large ad hoc wireless networks. *Proceeding of IEEE GLOBECOM 2000*, Nov. 2000.

[2] G. Pei, M. Gerla, and T. W. Chen. Fisheye state routing in mobile ad hoc networks. *Proceeding of ICDCS 2000 workshops*, Apr. 2000.

[3] C. Perkins and P. Bhagwat. Highly dynamic destination-sequenced distance-vector routing (DSDV) for mobile computers. *Proceeding of the ACM SIGCOMM'94*, Sep. 1994.

Chapter 2

COLLISION AVOIDANCE PROTOCOLS IN AD HOC NETWORKS *

J. J. Garcia-Luna-Aceves
Department of Computer Engineering
University of California at Santa Cruz
Santa Cruz, CA 95064, U.S.A.
jj@cse.ucsc.edu

Yu Wang
Department of Computer Engineering
University of California at Santa Cruz
Santa Cruz, CA 95064, U.S.A.
ywang@cse.ucsc.edu

Abstract We present an analytical model for the saturation throughput of sender-initiated collision avoidance protocols in multi-hop ad hoc networks with nodes randomly placed according to a two-dimensional Poisson distribution. We show that these protocols can accommodate much fewer competing nodes within a region in a network infested with hidden terminals than in those cases without hidden terminals or with just a few. These results are validated through computer simulations. We then introduce a framework to address the fairness problem inherent in ad hoc networks using IEEE 802.11 and propose a topology-aware fair access (TAFA) scheme to realize the framework. Simulation results show that TAFA can solve the fairness problem in UDP-based applications with negligible degradation in throughput, and the notorious problem of the starvation of flows in TCP-based applications while incurring only some throughput degradation.

Keywords: Collision avoidance, medium access control, ad hoc networks, fairness, IEEE 802.11, sender-initiated

---

*This work was supported in part by the Defense Advanced Research Projects Agency (DARPA) under Grant No. DAAD19-01-C-0026, the US Air Force/OSR under Grant No. F49620-00-1-0330 and the Jack Baskin Chair of Computer Engineering at UCSC.

Introduction

Wireless ad hoc networks have received increasing interest in recent years, because of their potential to be used in a variety of applications without the aid of any pre-existing network infrastructure.

Due to the scarce channel bandwidth available in ad hoc networks, the design of efficient and effective medium access control (MAC) protocols that regulate nodes' access to a shared channel has become the subject of active research in recent years. Many MAC protocols [1] [2] [3] [4] [5] have been proposed to mitigate the adverse effects of hidden terminals [6] through collision avoidance. Most collision avoidance schemes such as the carrier sense multiple access with collision avoidance (CSMA/CA) in the popular MAC protocols,IEEE 802.11 MAC protocol [2] are sender-initiated, including an exchange of short request-to-send (RTS) and clear-to-send (CTS) packets between a pair of sending and receiving nodes before the transmissions of the actual data packet and the optional acknowledgment packet.

In Section 2.1, we present an analytical modeling [7] to derive the saturation throughput of these sender-initiated collision avoidance protocols in multi-hop ad hoc networks with nodes randomly placed according to a two-dimensional Poisson distribution. We show that the sender-initiated collision-avoidance scheme achieves much higher throughput than the ideal carrier sense multiple access scheme with a separate channel for acknowledgments. More importantly, we show that the collision-avoidance scheme can accommodate much fewer competing nodes within a region in a network infested with hidden terminals than in a fully-connected network, if reasonable throughput is to be maintained. Simulations of the IEEE 802.11 MAC protocol and one of its variants validate the predictions made in the analysis.

The simulation results also reveal the fairness problem in IEEE 802.11 MAC protocol which refers to the severe throughput degradation of some nodes due to their unfavorable locations in the network and the commonly used binary exponential backoff (BEB) algorithm which always favors the node that last succeeds. This motivates the work presented in Section 2.2 in which we introduce a framework to address the fairness problem conclusively and propose a topology aware fair access (TAFA) scheme to realize the framework. Simulation results show that TAFA can solve the fairness problem in UDP-based applications with negligible degradation in throughput. It can also solve the notorious problem of the starvation of flows in TCP-based applications, while incurring only some throughput degradation. Hence, TAFA shows a much better overall tradeoff between throughput and fairness than other schemes previously proposed.

Section 2.3 concludes this chapter with directions for future work.

2.1 Performance of collision avoidance protocols

In Section 2.1.1, we present the analysis of the sender-initiated collision-avoidance scheme based on a four-way handshake and non-persistent carrier sensing, which can be also called the RTS/CTS-based scheme for the sake of simplicity. We first adopt a simple model in which nodes are randomly placed on a plane according to two-dimensional Poisson distribution with density λ. Varying λ has the effect of changing the congestion level within a region as well as the number of hidden terminals. In this model, it is also assumed that each node is ready to transmit independently in each time slot with probability p, where p is a protocol-dependent parameter. This model was first used by Takagi and Kleinrock [8] to derive the optimum transmission range of a node in a multi-hop wireless network, and was used subsequently by Wu and Varshney [9] to derive the throughputs of non-persistent CSMA and some variants of busy tone multiple access (BTMA) protocols [6]. Then we assume that both carrier sensing and collision avoidance work perfectly, that is, that nodes can accurately sense the channel busy or idle, and that the RTS/CTS scheme can avoid the transmission of data packets that collide with other packets at the receivers. The latter assumption can be called *perfect collision avoidance* and has been shown to be doable in the floor acquisition multiple access (FAMA) protocol [3]. Later we extend this model to take into account the possibility of data packets colliding with other transmissions, so that the model is also applicable to other MAC protocols, such as the popular IEEE 802.11 protocol, in which perfect collision avoidance is not strictly enforced.

In Section 2.1.2, we present numerical results from our analysis. We compare the performance of the sender-initiated collision avoidance scheme against the idealized non-persistent CSMA protocol in which a secondary channel is assumed to send acknowledgments in zero time and without collisions [6, 9], as the latter is the only protocol whose analysis for multi-hop ad hoc networks is available for comparison to date. It is shown that the RTS/CTS scheme can achieve far better throughput than the CSMA protocol, even when the overhead due to RTS/CTS exchange is high. The results illustrate the importance of enforcing collision avoidance in the RTS/CTS handshake.

However, the analytical results also indicate that the aggregate throughput of sender-initiated collision avoidance drops faster than that in a fully-connected network when the number of competing nodes within a region increases. This contrasts with conclusions drawn from the analysis of collision avoidance in fully-connected networks or networks with limited hidden terminals [3]. Our results show that hidden terminals degrade the performance of collision avoidance protocols beyond the basic effect of having a longer vulnerability period for RTSs. Hence, it follows that collision avoidance becomes more and more ineffective for a relatively crowded region with hidden terminals.

To validate the findings drawn from this analysis, in Section 2.1.3 we present simulations of the popular IEEE 802.11 MAC protocol. The simulation results clearly show that the IEEE 802.11 MAC protocol cannot ensure collision-free transmission of data packets, and that almost half of the data packets transmitted cannot be acknowledged due to collisions, even when the number of competing nodes in a neighborhood is only eight! However, the performance of the simulated IEEE 802.11 MAC protocol correlates well with what is predicted in the extended analysis, which takes into account the effect of data packet collisions and is used for the case when the number of competing nodes in a region is small. When the number of competing nodes in a region increases, the performance gap between IEEE 802.11 and the analysis decreases, which validates the statement that even a perfect collision-avoidance protocol loses its effectiveness gradually due to the random nature of the channel access and the limited information available to competing nodes.

The simulation results for the IEEE 802.11 protocol also show a larger variation in throughput than the predicted performance from the analytical model, which is due to its inherent fairness problems which motivates the second part of the work reported in this chapter.

2.1.1 Approximate Analysis

In this section, we derive the approximate throughput of a perfect collision avoidance protocol. In our network model, nodes are two-dimensionally Poisson distributed over a plane with density λ, i.e., the probability $p(i, S)$ of finding i nodes in an area of S is given by:

$$p(i, S) = \frac{(\lambda S)^i}{i!} e^{-\lambda S}.$$

Assume that each node has the same transmission and receiving range of R, and denote by N the average number of nodes within a circular region of radius R; therefore, we have $N = \lambda \pi R^2$.

To simplify our analysis, we assume that nodes operate in time-slotted mode. As prior results for CSMA and collision-avoidance protocols show [6], the performance of MAC protocols based on carrier sensing is much the same as the performance of their time-slotted counterparts in which the length of a time slot equals one propagation delay and the propagation delay is much smaller than the transmission time of data packets.

The length of each time slot is denoted by τ. Note that τ is not just the propagation delay, because it also includes the overhead due to the transmit-to-receive turn-around time, carrier sensing delay and processing time. In effect, τ represents the time required for all the nodes within the transmission range of a node to know the event that occurred τ seconds ago. The transmission times of RTS, CTS, data, and ACK packets are normalized with regard to τ, and are denoted by l_{rts}, l_{cts}, l_{data}, and l_{ack}, respectively. Thus, τ is also equivalent to 1 in later derivations. For the sake of simplicity, we also assume that all packet transmission times are multiples of the length of a time-slot.

We derive the protocol's throughput based on the heavy-traffic assumption, i.e., a node always has a packet in its buffer to be sent and the destination is chosen randomly from one of its neighbors. This is a fair assumption in ad hoc networks in which nodes are sending data and signaling packets continually. We also assume that a node is ready to transmit with probability p and not ready with probability $1-p$. Here p is a protocol-specific parameter that is slot independent. At the level of individual nodes, the probability of being ready to transmit may vary from time slot to slot, depending on the current states of both the channel and the node. However, because we are interested in deriving the average performance metrics instead of instantaneous or short-term metrics, the assumption of a fixed probability p may be considered as an averaged quantity that can still reasonably approximate the factual burstiness from a long-term point of view. In fact, this assumption is necessary to make the theoretical modeling tractable and has been extensively applied before [10] [8] [9]. For example, this model was used by Takagi and Kleinrock [8] to derive the optimal transmission range of a node in a multi-hop wireless network, and was used subsequently by Wu and Varshney [9] to derive the throughput of non-persistent CSMA and some variants of busy tone multiple access (BTMA) protocols [6].

It should also be noted that, even when a node is ready to transmit, it may transmit or not in the slot, depending on the collision avoidance and resolution schemes being used, as well as the channel's current state. Thus, we are more interested in the probability that a node transmits in a time slot, which is denoted by p'. Similar to the reasoning presented for p, we also assume that p' is independent at any time slot to make the analysis tractable. Given this simplification, p' can be defined to be

$$\begin{aligned} p' &= p \cdot \text{Prob.}\{\text{Channel is sensed idle in a slot}\} \\ &\approx p \cdot \Pi_I \end{aligned}$$

where Π_I is the limiting probability that the channel is in idle state, which we derive subsequently.

We are not interested in the exact relationship between p and p', and it is enough to obtain the range of values that p' can take, because the throughput of these protocols is mostly influenced by p'. To derive the rough relationship between p and p', we set up a channel model that includes two key simplifying assumptions.

First, we model the channel as a circular region in which there are some nodes. The nodes within the region can communicate with each other while they have weak interactions with nodes outside the region. *Weak interaction* means that the decision of inner nodes to transmit, defer and back off is almost

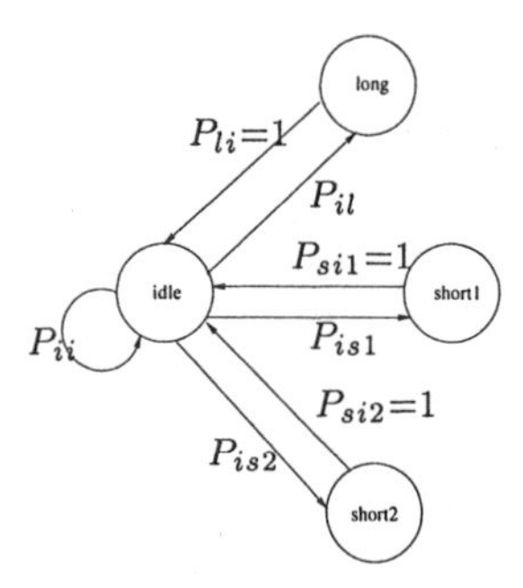

Figure 2.1. Markov chain model for the channel around a node

not affected by that of outer nodes and vice versa. Considering that nodes do not exchange status information explicitly (e.g., either defer due to collision avoidance or back off due to collision resolution), this assumption is reasonable and helps to simplify the model considerably. Thus, the channel's status is only decided by the successful and failed transmissions within the region.

Second, we still consider the failed handshakes initiated by nodes within the region to outside nodes, because this has a direct effect on the channel's usability for other nodes within the region. Though the radius of the circular region R' is unknown, it falls between $R/2$ and $2R$. This follows from noting that the maximal radius of a circular region in which all nodes are guaranteed to hear one another equals $R' = R/2$, and all the direct neighbors and hidden nodes are included into the region when $R' = 2R$. Thus, we obtain $R' = \alpha R$ where $0.5 \leq \alpha \leq 2$, and α needs to be estimated.

With the above assumptions, the channel can be modeled by a four-state Markov chain illustrated in Figure 2.1. The significance of the states of this Markov chain is the following:

- *Idle* is the state when the channel around node x is sensed idle, and obviously its duration is τ.

- *Long* is the state when a successful four-way handshake is done. For simplicity, we assume that the channel is in effect busy for the duration of the whole handshake, thus the busy time T_{long} is

$$\begin{aligned} T_{long} &= l_{rts} + \tau + l_{cts} + \tau + l_{data} + \tau + l_{ack} + \tau \\ &= l_{rts} + l_{cts} + l_{data} + l_{ack} + 4\tau. \end{aligned}$$

- *Short1* is the state when multiple nodes around the channel transmit RTS packets during the same time slot and their transmissions collide. The busy time of the channel T_{short1} is therefore

$$T_{short1} = l_{rts} + \tau.$$

- *Short2* is the state when one node around the channel initiates a failed handshake with a node outside the region. Even though a CTS packet may not be sent due to the collision of the sending node's RTS packet with other packets originated from nodes outside the region or due to the deferring of the receiving node to other nodes, those nodes overhearing the RTS as well as the sending node do not know if the handshake is successfully continued, until the time required for receiving a CTS packet elapses. Therefore the channel is in effect busy, i.e., unusable for all the nodes sharing the channel, for the time stated below:

$$\begin{aligned} T_{short2} &= l_{rts} + \tau + l_{cts} + \tau \\ &= l_{rts} + l_{cts} + 2\tau. \end{aligned}$$

Now we proceed to calculate the transition probabilities of the Markov chain.

In most collision avoidance schemes with non-persistent carrier sensing, no node is allowed to transmit immediately after the channel becomes idle, thus the transition probabilities from *long* to *idle*, from *short1* to *idle* and from *short2* to *idle* are all 1.

According to the Poisson distribution of the nodes, the probability of having i nodes within the receiving range R of x is $e^{-N}N^i/i!$, where $N = \lambda\pi R^2$. Therefore, the mean number of nodes that belong to the shared channel is $M = \lambda\pi R'^2 = \alpha^2 N$. Assuming that each node transmits independently, the probability that none of them transmits is $(1-p')^i$, where $(1-p')$ is the probability that a node does not transmit in a time slot. Because the transition probability P_{ii} from *idle* to *idle* is the probability that none of the neighboring nodes of x transmits in this slot, P_{ii} is given by

$$\begin{aligned} P_{ii} &= \sum_{i=0}^{\infty}(1-p')^i \frac{M^i}{i!} e^{-M} \\ &= \sum_{i=0}^{\infty} \frac{[(1-p')M]^i}{i!} e^{-(1-p')M} \cdot e^{-p'M} = e^{-p'M}. \end{aligned}$$

We average the probabilities over the number of interfering nodes in a region because of two reasons. First, it is much more tractable than the approach that conditions on the number of nodes, calculates the desired quantities, and then uses the Poisson distribution to obtain the average. Second, in our simulation experiments, we fix the number of competing nodes in a region (which is N) and then vary the location of the nodes to approximate the Poisson distribution, which is configurationally closer to our analytical model; the alternative would be to generate 2, 3, 4, ... nodes within one region, get the throughput for the individual configuration and then calculate the average, which is not practical.

Next we need to calculate the transition probability P_{il} from *idle* to *long*. If there are i nodes around node x, for such a transition to happen, one and only one node should be able to complete one successful four-way handshake while other nodes do not transmit. Let p_s denote the probability that a node begins a successful four-way handshake at each slot, we can then calculate P_{il} as follows:

$$\begin{aligned} P_{il} &= \sum_{i=1}^{\infty} i p_s (1-p')^{i-1} \frac{M^i}{i!} e^{-M} \\ &= \sum_{i=1}^{\infty} p_s (1-p')^{i-1} \frac{M^{i-1}}{(i-1)!} M e^{-M} \\ &= p_s M \sum_{i=0}^{\infty} \frac{[M(1-p')]^i}{i!} e^{-M(1-p'+p')} \\ &= p_s M e^{-p'M}. \end{aligned}$$

To obtain the above result, we use the fact that the distribution of the number of nodes within R' does not depend on the existence of node x, because of the memoryless property of the Poisson distribution. Up to this point, p_s is still an unknown quantity that we derive subsequently.

The transition probability from *idle* to *short1* is the probability that more than one node transmit RTS packets in the same slot; therefore, P_{is1} can be calculated as follows:

$$\begin{aligned} P_{is1} &= \sum_{i=2}^{\infty} [1-(1-p')^i - ip'(1-p')^{i-1}] \frac{M^i}{i!} e^{-M} \\ &= 1-(1+Mp')e^{-p'M}. \end{aligned}$$

Having calculated P_{ii}, P_{il} and P_{is1}, we can calculate P_{is2}, the transition probability from *idle* to *short2*

$$\begin{aligned} P_{is2} &= 1 - P_{ii} - P_{il} - P_{is1} \\ &= 1 - e^{-p'M} - p_s M e^{-p'M} - (1-(1+Mp')e^{-p'M}) \\ &= (p'-p_s) M e^{-p'M}. \end{aligned}$$

Let π_i, π_l, π_{s1} and π_{s2} denote the steady-state probabilities of states *idle*, *long*, *short1* and *short2*, respectively. From Figure 2.1, we have

$$\begin{aligned} \pi_i P_{ii} + \pi_l + \pi_{s1} + \pi_{s2} &= \pi_i \\ \pi_i P_{ii} + 1 - \pi_i &= \pi_i \\ \pi_i &= \frac{1}{2-P_{ii}} = \frac{1}{2-e^{-p'M}}. \end{aligned}$$

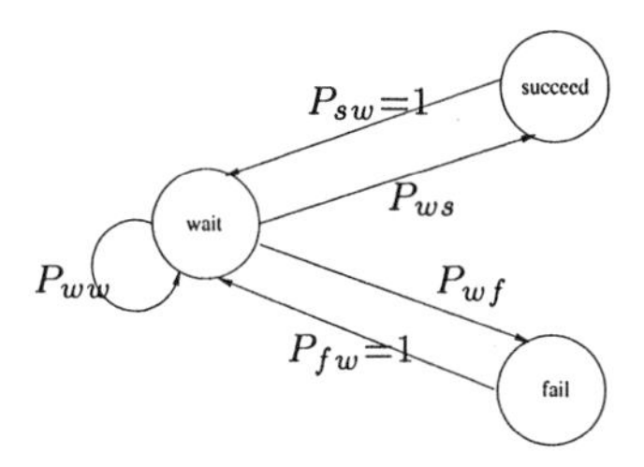

Figure 2.2. Markov chain model for a node

The limiting probability Π_I, i.e., the long run probability that the channel around node x is found idle, can be obtained by:

$$\Pi_I = \frac{\pi_i T_{idle}}{\pi_i T_{idle} + \pi_l T_{long} + \pi_{s1} T_{short1} + \pi_{s2} T_{short2}}.$$

Noting that $\pi_i P_{il} = \pi_l$, $\pi_i P_{is1} = \pi_{s1}$ and $\pi_i P_{is2} = \pi_{s2}$, we obtain

$$\begin{aligned}\Pi_I &= \frac{\pi_i T_{idle}}{\pi_i T_{idle} + \pi_i P_{il} T_{long} + \pi_i P_{is1} T_{short1} + \pi_i P_{is2} T_{short2}} \\ &= \frac{T_{idle}}{T_{idle} + P_{il} T_{long} + P_{is1} T_{short1} + P_{is2} T_{short2}}.\end{aligned}$$

The relationship between p' and p is then:

$$\begin{aligned}p' &= \frac{pT_{idle}}{T_{idle} + P_{il} T_{long} + P_{is1} T_{short1} + P_{is2} T_{short2}} \\ &= \frac{pT_{idle}}{T_{idle} + p_s M e^{-p'M} T_{long} + (1-(1+p'M)e^{-p'M}) T_{short1} + \cdots} \\ &\quad \overline{\cdots + (p'-p_s) M e^{-p'M} T_{short2}} \qquad (1.1)\end{aligned}$$

In the above equation, the probability that a node x starts successfully a four-way handshake in a time slot, p_s, is yet to be determined.

The states of a node x can be modeled by a three-state Markov chain, which is shown in Figure 2.2.

In Figure 2.2, *wait* is the state when the node defers for other nodes or backs off, *succeed* is the state when the node can complete a successful four-way handshake with other nodes, and *fail* is the state when the node initiates an unsuccessful handshake. For simplicity, we regard *succeed* and *fail* as the states when two different kinds of *virtual* packets are transmitted and their lengths are:

$$\begin{aligned}T_{succeed} &= T_{long} = l_{rts} + l_{cts} + l_{data} + l_{ack} + 4\tau \\ T_{fail} &= T_{short2} = l_{rts} + l_{cts} + 2\tau.\end{aligned}$$

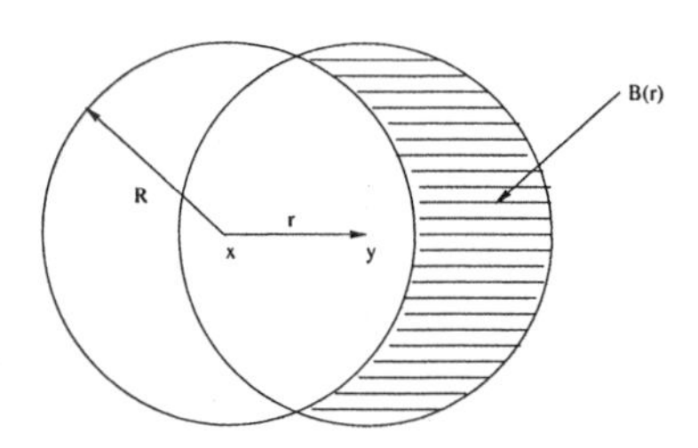

Figure 2.3. Illustration of "hidden" area

Obviously, the duration of a node in *wait* state T_{wait} is τ.

Because by assumption collision avoidance is enforced at each node, no node is allowed to transmit data packets continuously; therefore, the transition probabilities from *succeed* to *wait* and from *fail* to *wait* are both one.

To derive the transition probability P_{ws} from *wait* to *succeed*, we need to calculate the probability $P_{ws}(r)$ that node x successfully initiates a four-way handshake with node y at a given time slot when they are at a distance r apart. Before calculating $P_{ws}(r)$, we define $B(r)$ to be the area that is in the hearing region of node y but outside the hearing region of node x, i.e., the interfering region "hidden" from node x as the shaded area shown in Figure 2.3. $B(r)$ has been shown in [8] to be:

$$B(r) = \pi R^2 - 2R^2 q\left(\frac{r}{2R}\right) \tag{1.2}$$

where $q(t) = \arccos(t) - t\sqrt{1-t^2}$.

Then $P_{ws}(r)$ can be calculated as:

$$P_{ws}(r) = P_1 \cdot P_2 \cdot P_3 \cdot P_4(r)$$

where

$$\begin{aligned}
P_1 &= \text{Prob.}\{x \text{ transmits in a slot}\}, \\
P_2 &= \text{Prob.}\{y \text{ does not transmit in the time slot}\}, \\
P_3 &= \text{Prob.}\{\text{none of the terminals within } R \text{ of } x \text{ transmits in the same slot}\}, \\
P_4(r) &= \text{Prob.}\{\text{none of the terminals in } B(r) \text{ transmits for } (2l_{rts}+1) \text{ slots} \mid r\}.
\end{aligned}$$

The reason for the last term is that the vulnerable period for an RTS is only $2l_{rts} + 1$, and once the RTS is received successfully by the receiving node (which can then start sending the CTS), the probability of further collisions is assumed to be negligibly small.

Obviously, $P_1 = p'$ and $P_2 = (1-p')$. On the other hand, P_3 can be obtained by

$$\begin{aligned} P_3 &= \sum_{i=0}^{\infty}(1-p')^i \frac{(\lambda \pi R^2)^i}{i!} e^{-\lambda \pi R^2} \\ &= \sum_{i=0}^{\infty}(1-p')^i \frac{N^i}{i!} e^{-N} \\ &= e^{-p'N}. \end{aligned}$$

Similarly, the probability that none of the terminals in $B(r)$ transmits in a time slot is given by

$$\begin{aligned} p_4(r) &= \sum_{i=0}^{\infty}(1-p')^i \frac{(\lambda B(r))^i}{i!} e^{-\lambda B(r)} \\ &= e^{-p'\lambda B(r)}. \end{aligned}$$

Hence, $P_4(r)$ can be expressed as

$$\begin{aligned} P_4(r) &= (p_4(r))^{2l_{rts}+1} \\ &= e^{-p'\lambda B(r)(2l_{rts}+1)}. \end{aligned}$$

Given that each sending node chooses any one of its neighbors with equal probability and that the average number of nodes within a region of radius r is proportional to r^2, the probability density function of the distance r between node x and y is

$$f(r) = 2r, \quad 0 < r < 1.$$

where we have normalized r with regard to R by setting $R = 1$.

Now we can calculate P_{ws} as follows:

$$\begin{aligned} P_{ws} &= \int_0^1 2r P_{ws}(r) dr \\ &= 2p'(1-p')e^{-p'N} \int_0^1 r e^{-p'\lambda B(r)(2l_{rts}+1)} dr \\ &= 2p'(1-p')e^{-p'N} \int_0^1 r e^{-p'N[1-2q(r/2)/\pi](2l_{rts}+1)} dr. \end{aligned} \tag{1.3}$$

From the Markov chain shown in Figure 2.2, the transition probability P_{ww} that node x continues to stay in *wait* state in a slot is just $(1-p')e^{-p'N}$, i.e., node x does not initiate any transmission and there is no node around it initiating a transmission. Let π_s, π_w and π_f denote the steady-state probability of state *succeed*, *wait* and *fail*, respectively. From Figure 2.2, we have

$$\begin{aligned}\pi_w P_{ww} + \pi_s + \pi_f &= \pi_w \\ \pi_w P_{ww} + 1 - \pi_w &= \pi_w \\ \pi_w &= \frac{1}{2 - P_{ww}} = \frac{1}{2 - (1-p')e^{-p'N}}.\end{aligned}$$

Therefore, the steady-state probability of state *succeed*, π_s, can be calculated as:

$$\pi_s = \pi_w P_{ws} = \frac{P_{ws}}{2 - (1-p')e^{-p'N}} = p_s. \tag{1.4}$$

Equation (1.4) points out the fact that π_s is just the previous unknown quantity p_s in Equation (1.1). Combining Equations (1.1), (1.3) and (1.4) together, we get a complex relationship between p and p'. However, given p, p' can be computed easily with numerical methods.

Accordingly, the throughput Th is:

$$\begin{aligned}Th &= \frac{\pi_s \cdot l_{data}}{\pi_w T_w + \pi_s T_s + \pi_f T_f} \\ &= \frac{l_{data}\pi_s}{\tau\pi_w + (l_{rts} + l_{cts} + l_{data} + l_{ack} + 4\tau)\pi_s} \cdots \\ &\cdots \frac{}{+(l_{rts} + l_{cts} + 2\tau)(1 - \pi_w - \pi_s)}.\end{aligned} \tag{1.5}$$

From the formula used to calculate throughput, we can see that π_s and π_w, from which throughput is derived, are largely dependent on p' and not on p, which is the basis for our simplification of the modeling of the channel presented earlier.

To apply our analysis to MAC protocols in which perfect collision avoidance is not enforced, e.g., the IEEE 802.11 MAC protocol, we propose a simple though not rigorous extension of the analysis. We can add another state to the Markov chain for the node model (ref. Figure 2.2) whose duration is $l_{rts} + l_{cts} + l_{data} + 3\tau$. This is a *pseudo-succeed* state in which an RTS-CTS-data handshake takes place without acknowledgment coming back due to collisions, i.e., it is a state derived from the *succeed* state of the perfect collision avoidance protocol. We use an "imperfectness factor" β to model the deviatory behavior of the protocol, given that different MAC protocols may have different values of β. The transition probability from *wait* to the *pseudo-succeed* state is then βP_{ws}, and the transition probability from *wait* to *succeed* is $(1-\beta)P_{ws}$. Hence, the modified formula for throughput is simply:

$$Th = (1-\beta)l_{data}\pi_s[\tau\pi_w + (l_{rts}+l_{cts}+l_{data}+l_{ack}+4\tau)(1-\beta)\pi_s + (l_{rts}+l_{cts}+2\tau)(1-\pi_w-\pi_s) + (l_{rts}+l_{cts}+l_{data}+3\tau)\beta\pi_s]^{-1} \tag{1.6}$$

When the deviatory factor β equals zero, Equation (1.6) is reduced to Equation (1.5).

2.1.2 Numerical Results

In this section, we compare the throughput of the RTS/CTS scheme with a non-persistent CSMA protocol in which there is a separate channel over which acknowledgments are sent in zero time and without collisions. The performance of the latter protocol in multi-hop networks has been analyzed by Wu and Varshney [9] and we should note that, in practice, the performance of the CSMA protocol would be worse as both data packets and acknowledgments are transmitted in the same channel.

We present results when either relatively large data packets or relatively small data packets are sent. Let τ denote the duration of one time slot. RTS, CTS and ACK packets last 5τ. As to the size of data packets, we consider two cases. One case corresponds to a data packet that is much larger than the aggregate size of RTS, CTS and ACK packets. The other case corresponds to a data packet being only slightly larger than the aggregate size of RTS, CTS and ACK packets. In the latter case, which models networks in which radios have long turn-around times and data packets are short, it is doubtful whether a collision avoidance scheme should be employed at all, because it represents excessive overhead.

We first calculate throughput with different values of α, which we define as the ratio between the circular region including nodes affected by an RTS/CTS handshake and the largest possible circular region in which nodes are guaranteed to be connected with one another. We find that, though the relationship between the ready probability p and transmission-attempt probability p' under different values of α might be somewhat different, the throughput is largely unaffected by α, which is shown in Figure 2.4.[1] In Figure 2.4, N is the average number of nodes that compete against one another to access the shared channel. Thus, the burden of estimating α is relieved in our model, and we can focus on the case in which $\alpha = 1$ thereafter. However, as a side effect of not knowing the actual α that should be used, the relationship between p' and throughput may not agree with the simulations. However, for our purposes this is not a problem, because we are interested in the saturated throughput only.

[1]The curves for $N = 3$ with different values of α concentrates on the upper part of these figures while the ones for $N = 10$ on the lower part.

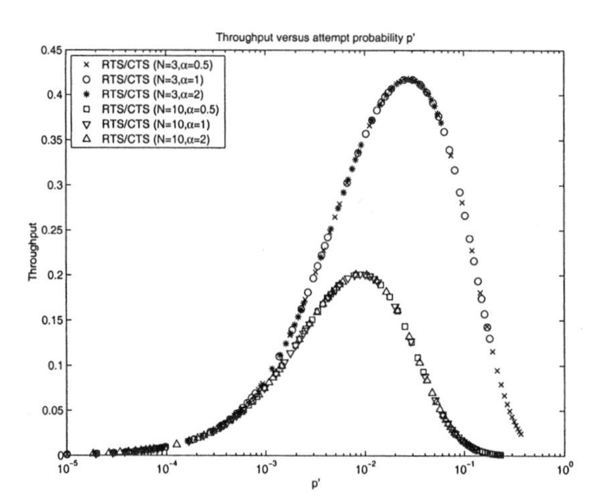

(a) long data packet: $l_{data} = 100\tau$

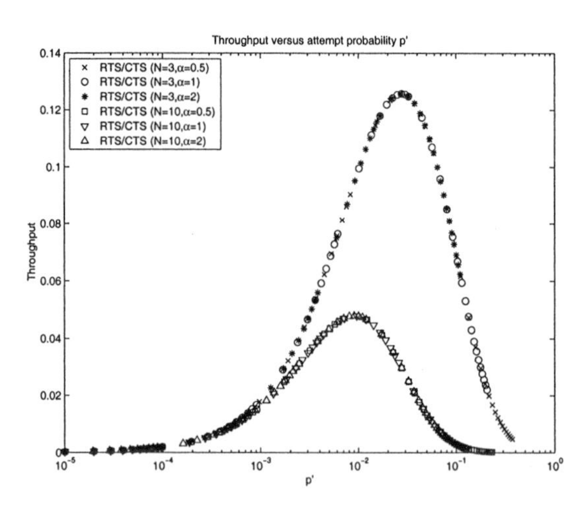

(b) short data packet: $l_{data} = 20\tau$

Figure 2.4. α's influence ($l_{rts} = l_{cts} = l_{ack} = 5\tau$)

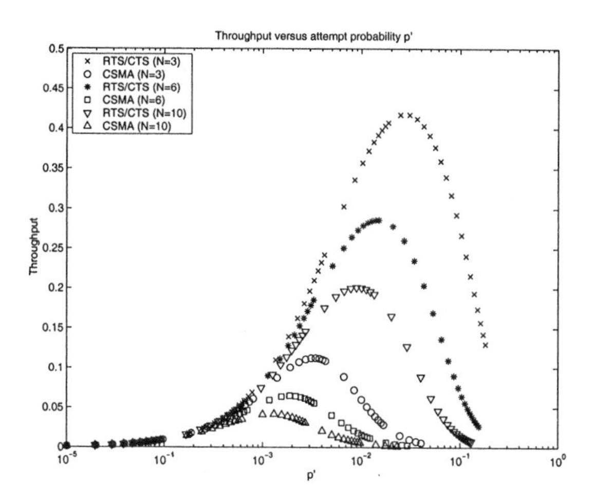

(a) long data packet: $l_{data} = 100\tau$

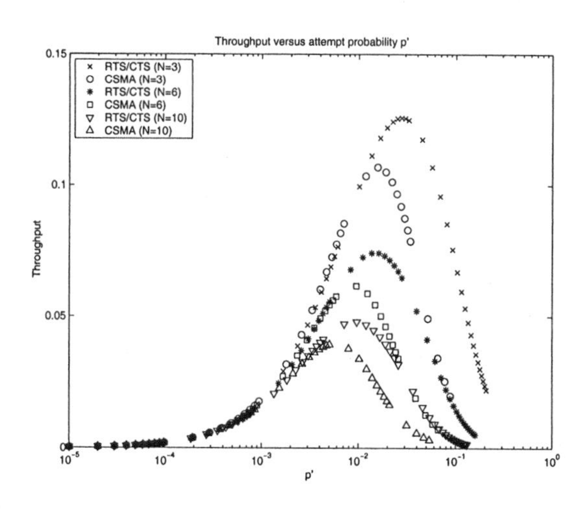

(b) short data packet: $l_{data} = 20\tau$

Figure 2.5. Throughput comparison ($l_{rts} = l_{cts} = l_{ack} = 5\tau$)

Figure 2.5 compares the throughput of collision avoidance against that of CSMA with different values of N and data packet lengths, and we can make the following observations from the above results.

When data packet is long, the throughput of CSMA is very low, even for the case in which only $N = 3$ nodes are competing for the shared channel. By comparison, the RTS/CTS scheme can achieve much higher throughput, even when the average number of competing nodes is 10. The reason is simple, the larger a data packet is, the worse the impact of hidden terminals is for that packet in CSMA, because the vulnerability period becomes twice the length of the data packet. With collision avoidance, the vulnerability period of a handshake is independent of the length of data packets, and in the worse case, equals twice the length of an RTS. When a data packet is not very long and the overhead of the collision avoidance and handshake seems to be rather high, collision avoidance can still achieve marginally better throughput than CSMA. We need to emphasize that the performance of the actual CSMA protocol would be much worse than the idealized model we have used for comparison purposes, because of the effect of acknowledgments.

Despite the advantage of collision avoidance, its throughput still degrades rapidly with the increase of N. This is also evident for low values of p' as shown in Figure 2.5. This is due the fact that nodes are spending much more time on collision avoidance and backoff. When N increases, p' decreases much slower to achieve optimum throughput, which already decreases. This shows that collision avoidance becomes more and more ineffective when the number of competing nodes within a region increases, even though these nodes are quite "polite" in their access to the shared channel. This is also different from a fully-connected network, in which the maximum throughput is largely indifferent to the number of nodes within a region [11].

Our results also reveal that hidden terminals degrade the performance of collision avoidance protocols beyond the basic effect of having a longer vulnerability period for RTSs. There is one dilemma here. On the one hand, it is very difficult to get all the competing nodes around one node coordinated well by probabilistic methods such as randomized backoff. Here the competing nodes refer to both one-hop and two-hop neighbors[2] of the node. In actual MAC protocols, the collisions of data packets may still occur and throughput degrades with increasing numbers of neighbors. On the other hand, even if all the competing nodes of one node defer their access for the node, the possible spatial reuse in multi-hop networks is greatly reduced and hence the maximum achievable throughput is reduced. This dilemma leads to the scalability problem of contention-based MAC protocols that occurs much earlier than people might expect, as the throughput is already quite meager when the average of competing nodes within a region (N) is only ten.

[2]Here we refer to those nodes that have at least one common neighbor with a node but are not direct neighbors of the node as the node's two-hop neighbors.

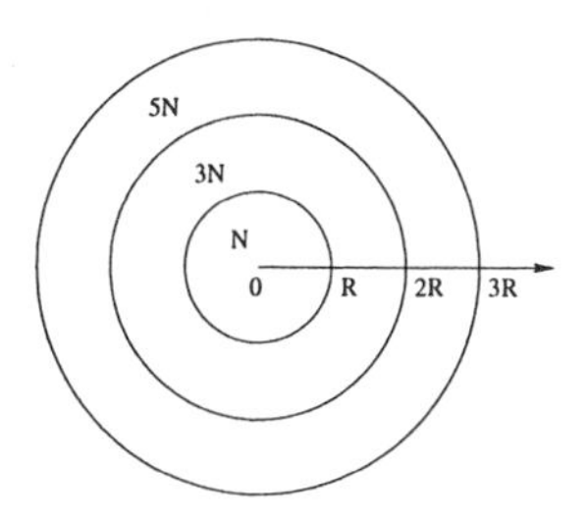

Figure 2.6. Network Model Illustration

2.1.3 Simulation Results

The numerical results in the previous section show that an RTS/CTS based access scheme outperforms CSMA, even when the overhead of RTS/CTS packets is comparable to the data packets to be transmitted if perfect collision avoidance can be achieved. In this section, we investigate the performance of the popular IEEE 802.11 DFWMAC protocol to validate the predictions made in the analysis.

We use GloMoSim 2.0 [12] as the network simulator. Direct sequence spread spectrum (DSSS) parameters are used throughout the simulations, which are shown in Table 2.1. The raw channel bit rate is 2Mbps. We use a uniform distribution to approximate the Poisson distribution used in our analytical model, because the latter is mainly used to facilitate our derivation of analytical results. In addition, it is simply impractical to generate 2, 3, 4, ... nodes within one region, get the throughput for the individual configuration and then calculate the average like what is required in the analytical model. In the network model used simulations, we place nodes in concentric circles or rings as illustrated in Figure 2.6. That is, given that a node's transmitting and receiving range is R and that there are on average N nodes within this circular region, we place N nodes in a circle of radius R, subject to a uniform distribution. Because there are on average 2^2N nodes within a circle of radius $2R$, we place $2^2N - N = 3N$ nodes outside the previous circle of radius R but inside the concentric circle of radius $2R$, i.e., the ring with radii R and $2R$, subject to the same uniform distribution. Then $3^2N - 2^2N = 5N$ nodes can be placed in an outer ring with radii $2R$ and $3R$.

Because it is impossible to generate the infinite network we assumed in our analysis in simulations, we just focus our attention on the performance of the innermost N nodes. Another reason is that it is more appropriate to investigate the performance of MAC schemes in a local neighborhood, rather than in the whole network, because totaling and averaging performance metrics such as throughput and delay with regard to all the nodes both in the center and at the edge of a network may lead to some askew results. For example, nodes at the

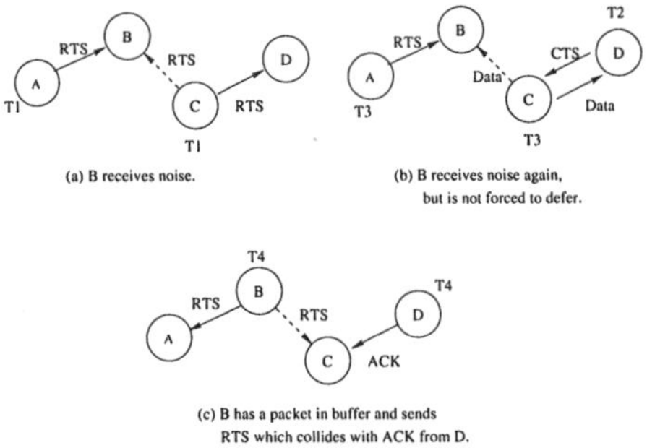

(a) B receives noise.

(b) B receives noise again, but is not forced to defer.

(c) B has a packet in buffer and sends RTS which collides with ACK from D.

Figure 2.7. Example of collisions with data packets in the IEEE 802.11 MAC Protocol

edge may have exceedingly high throughput due to much less contention and including them in the calculation would lead to higher than usual throughput. In our experiments, we find that nodes that are outside the concentric circles of radius $3R$ almost have no influence on the throughput of the innermost N nodes, i.e., boundary effects can be safely ignored when the circular network's radius is $3R$. Accordingly, we present only the results for a circular network of radius $3R$.

The backoff timer in the IEEE 802.11 MAC protocol is drawn from a uniform distribution whose upper bound varies according to the estimated contention level, i.e., a modified binary exponential backoff. Thus, p' takes on dynamic values rather than what we have assumed in the analytical model. Accordingly, we expect that the IEEE 802.11 MAC protocol will operate in a region, while our analysis gives only average performance. In addition, even in network topologies that satisfy the same uniform distribution, we can still get quite different results, which will be shown later.

As we have stated, the IEEE 802.11 MAC protocol cannot ensure collision-free transmission of data packets, even under the assumption of perfect carrier sensing and collision avoidance. There are two reasons for this. One is that the length of a CTS is shorter than that of an RTS, which has been shown to prevent some hidden nodes from backing off [3]. The other reason is that, when a node senses carrier in its surroundings, it does not defer access to the channel for a definite time (which is implicit in other protocols [3]) after the channel is clear. When the interfering node perceives the channel idle and a packet from the upper layer happens to arrive in its buffer, it may transmit immediately after the channel is idle for a DIFS (Distributed InterFrame Space) time, while in fact a data packet transmission may still be going on between another two nodes and collision will occur! This can be illustrated by the simple example shown in Figure 2.7.

In our simulation, each node has a constant-bit-rate (CBR) traffic generator with data packet size of 1460 bytes, and one of its neighbors is randomly chosen

Table 2.1. IEEE 802.11 protocol configuration parameters

RTS	*CTS*	*data*	*ACK*	*DIFS*	*SIFS*
20-byte	14-byte	1460-byte	14-byte	50μsec	10μsec

contention window	*slot time*	*sync. time*	*prop. delay*
31–1023	20μsec	192μsec	1μsec

Table 2.2. Equivalent configuration parameters for analytical model

	τ	l_{rts}	l_{cts}, l_{ack}	l_{data}
actual time	21μsec	272μsec	248μsec	6032μsec
normalized	1	13	12	287

as the destination for each packet generated. All nodes are always backloged. Considering the physical layer's synchronization time as well as propagation delay used in the simulation, the effective packet transmission times are shown in Table 2.1. For comparison purposes, we map these simulational parameters to equivalent parameters in our analytical model and they are shown in Table 2.2.

We run both analytical and simulation programs with $N = 3, 5$ and 8. Though we have not tried to characterize how the performance of the IEEE 802.11 MAC protocol is distributed in the region of values taken by p', we do have generated 50 random topologies that satisfy the uniform distribution and then get an average transmission probability and throughput for the N nodes in the innermost circle of radius R for each configuration. The results are shown in Figure 2.8, in which the centers of rectangles are the mean values of p' and throughput and their half widths and half heights are the variance of p' and throughput, respectively. These rectangles roughly describe the operating regions of IEEE 802.11 MAC protocol with the configurations we are using.

Figure 2.8 clearly shows that, IEEE 802.11 cannot achieve the performance predicted in the analysis of correct collision avoidance, but may well outperform the analysis with the same p' for some configurations, especially when N is small. On first thought, it may seem contrary to intuition, given that IEEE 802.11 cannot ensure collision-free data packet transmissions and should always perform worse than analysis results. In fact, the exceedingly high throughput is largely due to the unfairness of the binary exponential backoff (BEB) used in IEEE 802.11. In BEB, a node that just succeeds in sending a data packet

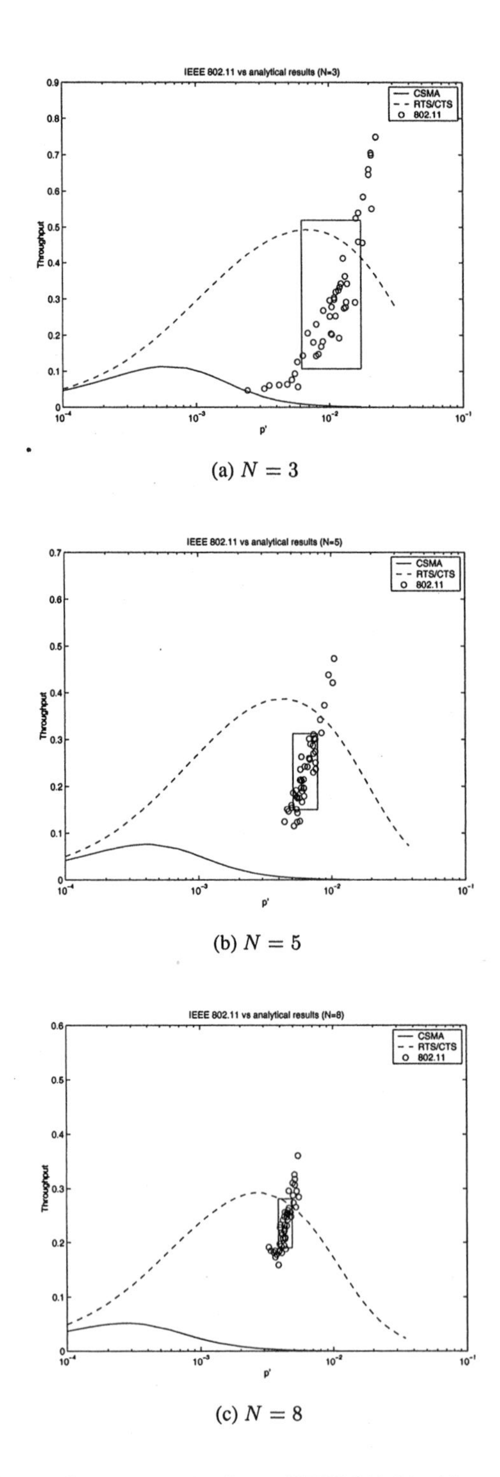

(a) $N = 3$

(b) $N = 5$

(c) $N = 8$

Figure 2.8. Performance comparison of IEEE 802.11 with analytical results

Table 2.3. Percentage of ACK timeout in BEB scheme

	N = 3	N = 5	N = 8
mean	0.29	0.39	0.44
std	0.17	0.10	0.06

resets its contention window to the minimum value, through which it may gain access to the channel again much earlier than other surrounding nodes. Thus, a node may monopolize the channel for a very long time during which there is no contention loss and throughput can be very high for a particular node, while other nodes suffer starvation. We also find that when N increases, the variance of p' and throughput becomes smaller. Thus, the fairness problem is less severe when there are more nodes competing in a shared channel.

Given that the IEEE 802.11 MAC protocol cannot ensure that data packets are transmitted free of collisions, its throughput can deviate much from what is predicted in the analysis. To demonstrate this, we also collect statistics about the number of transmitted RTS packets that will lead to ACK timeout due to collision of data packets as well as the total number of transmitted RTS packets that can lead to either an incomplete RTS-CTS-data handshake or a successful four-way handshake. Then we calculate the ratio of these two numbers and tabulate the results in Table 2.3. This table clearly shows that much of the precious channel resource is wasted in sending data packets that cannot be successfully delivered.

A close observation of Figure 2.8 also reveals that, the gap in maximum throughput between analytical and simulation results decreases when N increases. This can be explained as follows. When the number of direct competing nodes N increases, the number of indirect competing nodes (hidden terminals, $3N$ on average) also increases, which makes nodes implementing a perfect collision avoidance protocol spend much more time in deferring and backing off to coordinate with both one-hop and two-hop competing nodes to avoid collisions. Therefore, much of the gain of perfect collision avoidance is lost and possible spatial reuse is also reduced in congested area, which makes a perfect collision avoidance protocol work only marginally better than an imperfect one. This observation could not be predicted from previous analytical models or simulations focusing on fully-connected networks or networks with only a limited number of hidden terminals [11] [10] [13].

The percentage shown in Table 2.3 is in fact the β in our extended analysis to explain the deviatory behavior of MAC protocols that do not have perfect collision avoidance. Using these values, we compare the performance of the

IEEE 802.11 protocol with that of the adjusted analysis obtained from Equation (1.6), and show the results in Figure 2.9. In Figure 2.9, we only show the results for small values of N as it is not quite meaningful to do the adjustment for large values of N due the reason stated above. Figure 2.9 shows that the extended analysis is a rather good approximation of the actual performance of the IEEE 802.11 protocol though the latter has larger variation in throughput (possibly due to its inherent fairness problems).

2.2 Framework and Mechanisms for Fair Access in IEEE 802.11

As we have stated, the fairness problem is due to some nodes' unfavorable location in the network and the commonly used binary exponential backoff (BEB) aggravates this problem. The fairness problem is not new and there is already some work done on it. The work so far can be roughly categorized into two classes. In the first class, the goal is to achieve max-min fairness [14] [15] [16] by reducing the ratio between maximum throughput and minimum throughput of flows, either at a node's level or at a flow's level. In the second class, the approach used in fair queuing for wireline networks is adapted to multi-hop ad hoc networks taking into account location dependent contention [17] [18] [19] [20] [21] and flow contention graphs are used extensively in the schemes in the second class to model the contention among nodes. Figure 2.10 shows an example of how this is done. Any two flows with adjacent vertices in the flow contention graph should not be scheduled to transmit at the same time. Despite the differences of backoff algorithms and information exchange among these schemes, the underlying channel access scheme remains largely the basic sender-initiated collision avoidance handshake, which can be less effective than a receiver-initiated scheme when a receiver has better knowledge of the contention around itself than the sender.

Based on this key observation, in our earlier work [22], we proposed a hybrid channel access scheme that combines both sender-initiated and receiver-initiated collision avoidance handshake to address the fairness problem. The attractiveness of this approach is that it is compatible with the IEEE 802.11 framework and involves only some additional queue management and bookkeeping work. However, this recent work has shown that, despite its simplicity, it is not very effective for TCP-based flows and that more information exchange among nodes is necessary to solve the fairness problem conclusively. This motivates us to further our work on a framework to address the fairness problem in a systematic way. In Section 2.2.1, we identify several key components that constitutes our fairness framework and explain the rationale for their necessity. In Section 2.2.2, we propose new algorithms to realize the fairness framework.

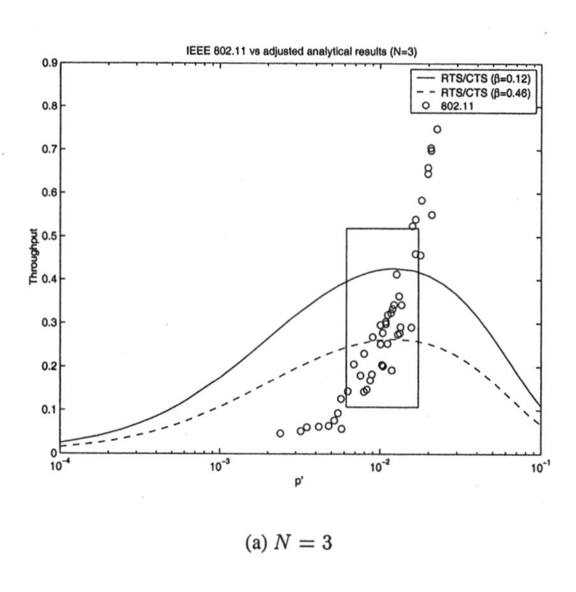

(a) $N = 3$

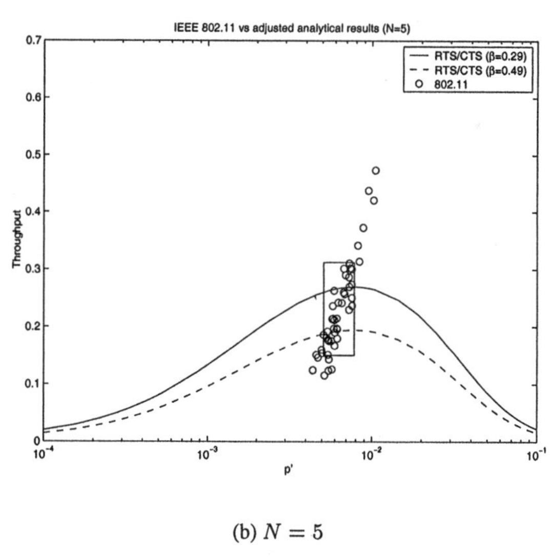

(b) $N = 5$

Figure 2.9. Performance comparison of IEEE 802.11 with adjusted analytical results

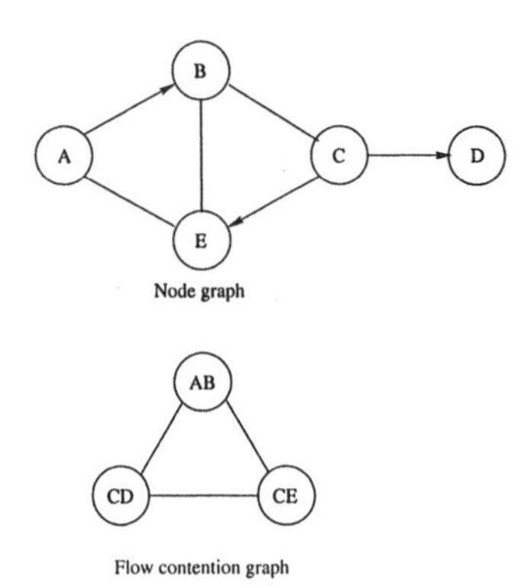

Figure 2.10. A simple network: node graph and flow contention graph

The resulting scheme, which we simply call topology aware fair access (TAFA) is evaluated in Section 2.2.3 through computer simulations. The performance of TAFA is compared with that of the original IEEE 802.11 MAC protocol and the hybrid channel access scheme proposed in [22] for both UDP- and TCP-based traffic. Simulation results show that TAFA can solve the fairness problem in UDP-based applications with negligible degradation in throughput. It can also solve the notorious problem of starvation of flows in TCP-based applications, despite some moderate degradation in throughput. Hence, TAFA shows a much better overall tradeoff between throughput and fairness than the other schemes investigated.

2.2.1 The Fairness Framework

In this section, we describe a framework for achieving better fairness consisting of four key components:

- Exchange of flow information among nodes;
- Adaptive backoff algorithm that is as stable as binary exponential backoff (BEB) but does not have the inherent deficiency of aggravating the fairness problem;
- Switching sender-initiated and receiver-initiated scheme as appropriate;
- Dealing with two-way flows.

The need for the exchange and maintenance of flow contention information can be illustrated by a simple example with the network configuration 4-8 shown in Figure 2.11. In Figure 2.11, a dashed line means that two nodes can hear each other's transmissions and an arrow indicates an active flow between two nodes. Nodes without any line in-between are hidden from each other. For configuration 4-8, node 2 knows that both node 0 and node 3 are sending nodes. However, if node 2 does not explicitly tell both node 0 and node 3 about the

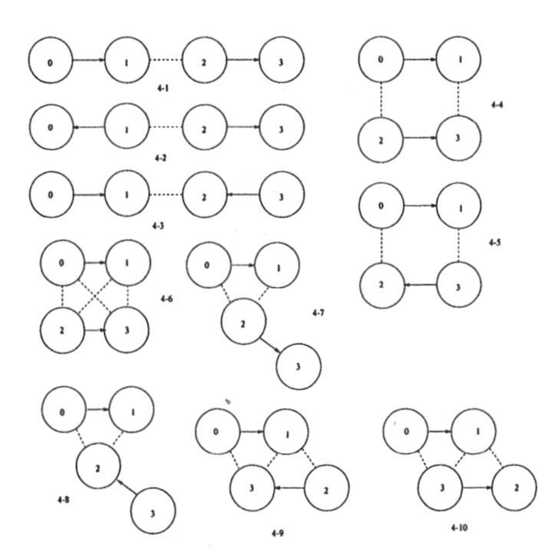

Figure 2.11. Network configurations with two competing flows

existence of each other, the handshake between node 0 and node 1 will tend to dominate the channel, because node 3's transmissions will mostly collide with either node 0 or node 1's transmissions at node 2, and both node 0 and node 1 may incorrectly perceive that node 0 and node 1 are the only active nodes in the network. Even though they may receive node 2's packets sporadically and make some ad hoc adjustment, without a systematic way to obtain flow information, the fairness problem cannot be solved conclusively.

The second component of our framework is an adaptive backoff scheme which is mandatory because the existing binary exponential backoff can aggravate the fairness problem as shown extensively in the literature [1] [14] [15]. Nodes should decide their channel access based on the information of competing flows gathered through the first component.

The third component of our framework is a hybrid channel access scheme that combines both sender-initiated and receiver-initiated collision handshake. This is largely due to the advantage of distributing the burden of initiating collision avoidance handshake between a pair of sending and receiving nodes depending on the different degrees of contention they experience. For example, in the network configuration 4-1 shown in Figure 2.11, the flow from node 0 to node 1 will suffer severe throughput degradation if no proper action is taken, because RTS from node 2 can always be received by node 3 successfully while node 0's RTS collides with node 2's transmissions at node 1 most of the time. In this case, if the collision avoidance is initiated by node 1, which transmits CTS to node 0 directly, then the channel bandwidth will be shared between these two flows more evenly, because node 1 and node 2 are direct neighbors and it is easier for them to coordinate their access to the channel.

The fourth component of our framework is a key contribution of the framework and consists of dealing with two-way traffic in which there are one data flow and one acknowledgment flow between two nodes, as is the case in most TCP-based flows. In such cases, usually one node cannot continue sending data packets, unless it receives application level acknowledgment packets from the other node. Though viewed from a traditional MAC's perspective they are separate flows, the performance of these two flows is coupled and they should compete as a collective entity rather than do so separately. Fairness for such cases is only touched upon in [22] and has not been addressed adequately in the literature, because most of the performance evaluation of fair MAC schemes so far has been done with constant bit rate (CBR) like traffic. The information about whether a flow is one-way or has a reverse flow can be conveyed from the application down to the MAC layer through some interface, which is not discussed here. We believe that such information and hence the required special processing are necessary to achieve the desired fairness goal.

2.2.2 Topology-Aware Fair Access

The topology aware fair access (TAFA) scheme is a realization of the fairness framework described previously, and consists of four parts corresponding to the four components in the framework.

2.2.2.1 Exchange and Maintenance of Flow Information. Each node maintains a flow table and each entry in the table contains the following information about a flow: *source address*, *destination address*, *service tag*, *direct flag* and *position flag*.

The service tag is used to measure how much channel resource the flow has received. Though there can be several ways to calculate the service tag, we use a simple one, which consists of the number of bytes that have been sent by the sender and acknowledged by the receiver. The service tag is updated by the sender when it receives an acknowledgment from the receiver and updated information is propagated to other nodes through subsequent packet transmissions.

The direct flag is used to indicate whether the flow is known directly through listening to the channel or indirectly through flow advertisements from other nodes. For example, in the network configuration 4-8 shown in Figure 2.11, node 3 cannot know the flow from node 0 to node 1 directly and has to rely on node 2 to advertise that flow to it. In this case, the flow from node 0 to node 1 is recorded as indirect in node 3's flow table and node 3 does not advertise the indirect flow.

The position flag is used to indicate whether a flow is original, a derivative, or not applicable to either case. This flag is used to handle two-way traffic. For example, in some TCP-based applications, one end of the connection cannot continue sending packets, unless it receives a TCP acknowledgment from the other end. The MAC protocol cannot just treat the data flow and the acknowledgment flow as separate flows. Due to the asymmetry of most connections, i.e., a data flow usually generates much more traffic than the corresponding acknowledgment flow, trying to equate the channel utilization for both flows would lead to throughput degradation. So it is important to use the position flag to indicate whether the flow is *original* (data flow) or *derivative* (acknowledgment flow) and the service tag of a derivative flow should be adjusted according to that of the corresponding original flow.

In this scheme, an RTS or a CTS only carry the information about the current flow (from the sender to the receiver) to reduce the fixed overhead that exists whether fairness is desired or not. Because the source and destination of a flow is self evident and a direct flag is not necessary, the extra information included in the RTS and CTS is just the service tag and the position flag of the flow. A receiver just copies the service tag in an RTS to its outgoing CTS, so that the neighbors of the receiver can also know the service tag of the ongoing flow. On the other hand, data packets and ACKs carry extra information about other flows maintained by the node if necessary. The rationale for treating these control packets differently is that the size of an RTS and a CTS can be fixed and nodes can get the duration information of the subsequent handshake from the network allocation vector (NAV) embedded in all packets. Because data packets are of varying size, it is acceptable for them to carry a bit more information. An ACK should also carry some extra flow information, otherwise those nodes that are neighbors of the node sending the ACK will never get any information about the flows around the node if the node does not send any data packet.

Specifically, to reduce the overhead incurred in the flow information exchange, nodes advertise only one flow at a time in the data or ACK packets they transmit, and one flow is chosen from the node's flow table in a round-robin way. As stated earlier, they only advertise flows that they know directly through receiving transmissions from either the sender or the receiver of the flow, rather than through the advertisement by other nodes. This avoids building up all the flows' information in a node which is unnecessary because channel access should be a local decision based only on the information of flows competing directly to avoid the complexity of making global decisions which is not what MAC layer should consider. Besides, nodes can obtain the updates of neighbor flows more quickly because only such flows are advertised. Flow information adversed in data and ACK packets includes only the source address, destination address and service tag.

Through the advertisement of flows, a node comes to know the other flows that may be competing with itself, gathers neighborhood topology information naturally, and adjusts its channel access accordingly.

2.2.2.2 Flow Aware Backoff Algorithm. In this scheme, each node also maintains two flags: *MyFlow* and *OtherFlow*. When a node receives the acknowledgment for its data packet, it updates its service tag and sets *MyFlow* true. When a node receives updated and greater service tag for other flows, it sets *OtherFlow* true. These two flags are used for a node to decide its contention window (CW), which is the upper bound of the uniform distribution from which a backoff timer is generated.

Unlike other schemes that deviate significantly from the binary exponential backoff (BEB) used in the IEEE 802.11 MAC protocol, we adopt BEB's basic idea of quick contention resolution and robustness and the resulting backoff algorithm is shown in Figure 2.12 in pseudo-code. Lines from 1 through 7 deal with the case when the node is the sender the flow with the minimum service tag. If neither the flow nor any other flow progresses (lines 2–3), then it means that some other nodes may also perceive that they have the minimum flows and it is important for the node to double its contention window (CW) for quick contention resolution. If any other flow progresses (lines 4–5), then the node should keep its current CW lest it may cause collisions by decreasing the CW and suffer unfairness by increasing the current CW, because it is already lagging behind other flows. If this flow has already made progress (lines 6–7), then it is safe to set its CW to the minimum value, because there is no perceived immediate contention from other flows. Lines from 9 through 17 deal with the case when the node does not have the minimum flow. If neither my flow nor other flow progresses (lines 9–10), it is important to double the CW for quick contention resolution. If only other flows make progress (lines 11–12), then it is adequate to keep the current CW, because the node does not require immediate access to the channel. However, if only my flow progresses (lines 13-14), then it means that the node is too aggressive in its channel access and should double its CW to yield the channel access to the other nodes that have minimum flows. If both my flow and other flow progress, then the node can reset the CW to the minimum value to avoid too much time spent in backoff. At last, in line 18, both *MyFlow* and *OtherFlow* are cleared and the backoff algorithm will be adapted again to any future change made to these two flags.

2.2.2.3 Topology-Aware Hybrid Collision Avoidance Handshake. As we have discussed, sometimes receiver-initiated collision avoidance can be more effective than sender-initiated and a combination of both is shown to yield quite satisfactory results when used to address the fairness problem [22].

To put our scheme in perspective, we give a brief review of the hybrid channel access scheme proposed in [22]. To maintain its compatibility with IEEE 802.11, the hybrid scheme does not introduce new types of control packets. Instead, a CTS packet is reused as the polling packet. Hence, the receiver-initiated

```
1: if (My flow has the min service tag in my flow table) {
2:     if (!MyFlow && !OtherFlow)
3:             Double contention window;
4:     else if (OtherFlow)
5:             Keep current contention window;
6:     else if (MyFlow)
7:             Reset contention window to minimum;
8: } else {
9:     if (!MyFlow && !OtherFlow)
10:            Double contention window;
11:    else if (OtherFlow && !MyFlow)
12:            Keep current contention window;
13:    else if (MyFlow && !OtherFlow)
14:            Double contention window;
15:    else if (MyFlow && OtherFlow)
16:            Reset contention window to minimum;
17: }
18: Clear MyFlow and OtherFlow.
```

Figure 2.12. The adaptive backoff algorithm

collision avoidance handshake just includes a three-way CTS-data-ACK exchange between polling and polled nodes.

Nodes implementing the hybrid scheme alternate in two modes: Sender-initiated (SI) and receive-initiated (RI). Nodes by default stay in the SI mode and use the usual four-way RTS-CTS-data-ACK handshake of IEEE 802.11. When a node transmits its RTS and fails to get the CTS from the intended receiver for several times, it sets the RI flag in the header of subsequent packets it sends and invites the receiver to start a receiver-initiated handshake. After receiving such packets, the receiver will confirm the sender with the RI flag also set in its reply if the receiver also implements the hybrid scheme. Upon receiving this confirmation, the sender will not transmit RTS packets to the receiver further. Instead, it just waits for the receiver to initiate a CTS-data-ACK handshake to itself, thus avoiding aggravating the contention for the channel. At this time, both nodes are engaged in the RI mode. A sender renews its RI request by setting RI flag continuously in the packets it sends out and cancels its request by clearing the flag, for example, when it has no more packet for the receiver.

The criterion to trigger the receiver-initiated handshake in [22] is that a node sets the RI request flag in its packets after it has sent the same RTS packet for more than one half of the times allowed in the IEEE 802.11 MAC protocol and receives no response from the intended receiver. The problem with this approach is that the receiver can hardly get any RTS sometimes due to high contention around it and hence receiver-initiated handshake cannot be triggered. This phenomenon is especially conspicuous for a two-way TCP connection, which consists of one data flow and one acknowledgment flow, because a pair of nodes may take turns to grab the channel, while other less privileged nodes may defer their access to the channel further due to the flow control and congestion avoidance functions in TCP.

To address the above problem, we propose a topology-aware scheme to switching between sender-initiated and receiver-initiated handshake. The basic idea is to make nodes that are closer to the contention initiate the handshake. To facilitate the description of the algorithm, some notations are used as shown in Table 2.4. Two flows are called dependent if they need to take turns to proceed, like a data flow and an acknowledgment flow in most TCP-based flows. That is why the position flag is exchanged and recorded in a node's flow table.

Figure 2.13 shows the criteria to switch between sender-initiated and receiver-initiated handshake. Similar to the algorithm shown in Figure 2.12, lines from 1 through 7 deal with the case when the node is the sender of the flow with the minimum service tag. If there is any independent flow in this node's table (lines 2–6), then the node needs to differentiate between two cases. If either the sender or the receiver of the independent flow which has the minimum service tag is this node's neighbor (line 3), then the usual sender-initiated handshake is used (line 4). Otherwise, it is possible that the receiver of the node is closer to

Table 2.4. Notations used in the hybrid scheme

V	The node applying the algorithm
f_m	The flow with the minimum service tag among all the flows in node V's flow table
f_{mi}	The flow with the minimum service tag among all the flows in node V's flow table that are not dependent on flows originating from node V
$S(f)$	Sender of flow f
$R(f)$	Receiver of flow f
$N(V)$	Node V's neighbors

either the sender or the receiver of that independent flow and it is more appropriate for the node to ask its receiver to use receiver-initiated handshake (line 6). In this way, the node and its receiver may compete for the channel more effectively. If the node does not have the minimum flow (lines 8–13), it should find the minimum flow in its flow table first. If the node is the receiver of the minimum flow or either the sender or the receiver of the minimum flow is its neighbor, then it just stays in the SI mode (lines 9–11). Otherwise, it means that the receiver of its flow may be closer to the nodes having the minimum flow, and then the node asks its receiver to enter the RI mode (line 12) with the hope that its receiver may compete for the channel more effectively than itself.

2.2.2.4 Dealing with Two-Way Flows.

Two-way flows require special processing as discussed before. We describe some necessary changes to the algorithms discussed in the previous subsections.

For an original flow and a derivative flow to compete for the channel effectively, the key idea is that the service tags for these flows in the participating nodes' flow tables should have correct relationship, i.e., if $T(f_1) \leq T(f_2)$ in one node's flow table, then it should be the same in the other node's flow table, so that nodes can make correct decisions in the backoff algorithm and the switch between sender-initiated and receiver-initiated handshake. It does not matter even if there are some discrepancies about the service tags of these flows maintained individually by each node.

In dealing with two-way flows, it is important to differentiate between original and derivative flows: The original flow is the one from the node that initiates the connection to the other node that acknowledges the connection. Then the required special processing can be summarized in two rules.

```
1:  if (S(f_m) is V) { // Our flow has the min service tag
2:          if (∃f_mi) { // If an independent flow is found;
3:                  if (S(f_mi) ∈ N(V) or R(f_mi) ∈ N(V))
4:                          Sender-initiated;
5:                  else
6:                          Receiver-initiated;
7:          } else Sender-initiated;
8:  } else { // Some other flow has the min service tag
9:          if (V is R(f_m) or S(f_m) ∈ N(V)
10:                         or R(f_m) ∈ N(V))
11:                 Sender-initiated;
12:         else    Receiver-initiated;
13: }
```

Figure 2.13. The criteria to choose sender-initiated or receiver-initiated handshake

Rule 1: When a node that initiates the original flow receives a packet from the corresponding derivative flow, it sets the service tag for the derivative flow (maintained in its flow table) to be the service tag of the original flow plus the size of the acknowledged data packet measured in bytes. It does not change the *OtherFlow* flag because in fact the derivative flow is not an independent flow.

Rule 2: When a node that is the sender of a derivative flow receives a packet from the corresponding original flow, it updates the service tags for both flows in its table as follows. Let f_o denote the received service tag of the original flow and f_d the current service tag of the derivative flow in its table. Then the new service tags for the original flow (f_{no}) and the derivative flow (f_{nd}) are: $f_{nd} = f_o$ and $f_{no} = f_o + f_d$. In this case, the node does not change *MyFlow* flag because the node itself is in effect not making any real progress.

Figure 2.14 shows the algorithm when the above two rules are applied.

How to apply these rules are better illustrated by the example shown in Table 2.5. In this example, the packet from the original flow (0→1) has a size of 100 bytes, and the packet from the derivative flow (1→0) has a size of 4 bytes. *My* and *Other* are the short names for the *MyFlow* and *OtherFlow* flags. It is clear that these two rules make sure that the service tags of these two flows have the correct relationship in either node's table even if they are not up-to-date.

2.2.3 Simulation Results

In our simulations, we focus on how two competing flows share the available channel resource in a few simple network configurations. These configurations are shown in Figure 2.11. Despite the simpleness of these configurations, it is

1: **if** (My flow is original) {
2: **if** (Receive a data packet T_d from the derivative flow)
3: $f_{nd} = f_o + T_d$;
4: } **else if** (My flow is derivative) {
5: **if** (Receive a data packet from the original flow)
6: $f_{nd} = f_o$; $f_{no} = f_o + f_d$;
7: }

Figure 2.14. Special tag processing for two-way flows

Table 2.5. An example of two-way flow processing

Time	*Event*	*Node 0*				*Node 1*			
		0→1	1→0	*My*	*Other*	0→1	1→0	*My*	*Other*
t_0	initialization	0	0	-	-	0	0	-	-
t_1	0 sends data, 1 acks	100	0	1	-	0	0	-	-
t_2	1 sends data, 0 acks	100	104	1	-	0	4	1	-
t_3	0 sends data	100	104	1	-	100	4	1	-
t_4	1 acks	200	104	1	-	104	100	1	-

Table 2.6. IEEE 802.11 and TAFA specific configuration parameters

	RTS	*CTS*	*data header*	*ACK*
802.11	20-byte	14-byte	28-byte	14-byte
TAFA	28-byte	22-byte	48-byte	34-byte

interesting to note that the fair schemes [1] [14] [15] [22] proposed so far have not addressed all the fairness problems in these network configurations when flows are either UDP- or TCP-based.

We use GloMoSim 2.0 [12] as the network simulator and our implementation of the new scheme (TAFA) is based on the IEEE 802.11 MAC protocol. Below are some details of the implementation. For RTS/CTS, we add three fields: service tag (4 bytes), position flag (2 byte) of current flow and receiver-initiated (RI) flag (2 byte). Though 1 byte should be enough for any of these flags, we choose larger size to allow easy extensions in the future if any. For data and ACK, in addition to the above three fields, they also include an advertisement about a flow from its flow table which includes three fields: source address (4 bytes), destination address (4 bytes) and service tag (4 bytes). In our implementation, a node indicates explicitly its originality in the RTS/data packets it sends out if applicable. All these constitute the fixed packet overhead in using the new scheme.

For IEEE 802.11, direct sequence spread spectrum (DSSS) parameters are used throughout the simulations. Most of the parameters remain the same as shown in Table 2.1 and protocol specific configuration parameters are shown in Table 2.6.

We investigate the performance of the IEEE 802.11 MAC protocol, the hybrid channel access scheme (for simplicity, it is simply called *Hybrid* thereafter) and the TAFA scheme under both UDP- and TCP-based traffic. In the first set of the simulation experiments, there are two competing UDP-based flows. For each flow, one node keeps sending data packets to the other at a constant bit rate, such that the sending queue is always non-empty. UDP is the underlying transport layer, thus no acknowledgment packet is sent back to the initiating node. We ran each configuration five times with different seed numbers and with a duration of 30 seconds. If the standard deviation of throughput is within 10% of the mean throughput, we show mean values only. Otherwise, we show both mean and standard deviation of the throughput. Table 2.7 shows the configurations when the IEEE 802.11 MAC protocol has fairness problem or there is some difference among these schemes. The "-" sign in the rows for the hybrid scheme indicates that receiver-initiated handshake is not triggered at all.

Table 2.7. Throughput comparison for the IEEE 802.11, the hybrid scheme and TAFA – two CBR flows (throughput measured in kbps)

Config #	Scheme	Flow #	Throughput	Flow #	Throughput	Aggregate
4-1	802.11	$0 \to 1$	83.4	$2 \to 3$	1500	1580
	Hybrid	$0 \to 1$	369	$2 \to 3$	1230	1600
	TAFA	$0 \to 1$	771	$2 \to 3$	778	1550
4-2	802.11	$1 \to 0$	820	$2 \to 3$	814	1630
	Hybrid	$1 \to 0$	-	$2 \to 3$	-	-
	TAFA	$0 \to 1$	769	$3 \to 2$	769	1540
4-3	802.11	$0 \to 1$	688	$3 \to 2$	709	1400
	Hybrid	$0 \to 1$	665	$3 \to 2$	643	1310
	TAFA	$0 \to 1$	683	$3 \to 2$	656	1340
4-7	802.11	$0 \to 1$	783	$2 \to 3$	824	1610
	Hybrid	$0 \to 1$	-	$2 \to 3$	-	-
	TAFA	$0 \to 1$	764	$2 \to 3$	764	1530
4-8	802.11	$0 \to 1$	1550	$3 \to 2$	28	1580
	Hybrid	$0 \to 1$	1280	$3 \to 2$	319	1600
	TAFA	$0 \to 1$	773	$3 \to 2$	805	1580
4-9	802.11	$0 \to 1$	734	$2 \to 3$	809	1540
	Hybrid	$0 \to 1$	815	$2 \to 3$	742	1560
	TAFA	$0 \to 1$	681	$2 \to 3$	676	1360

It can be seen that in some configurations such as 4-3 and 4-9 when the existing IEEE 802.11 MAC protocol works well, both the hybrid scheme and TAFA are unnecessary. Still, throughput degradation in TAFA is negligible. On the other hand, in configurations such as 4-1 and 4-8 where serious fairness problems occur in 802.11, TAFA shows superior performance to the other two schemes.

In the second set of simulation experiments, there are two competing TCP-based flows. We use the FTP/Generic application provided in GloMoSim, in which a client simply sends data packets to a server without the server sending any control information back to the client other than the acknowledgment packets required by TCP. Whenever a packet indicates success of delivery by the transport layer (TCP), the client sends the next data packet. It should be noted that the acknowledgment packet from TCP is still regarded as a normal data packet from the view of traditional MAC layer, which does not provide special processing for two-way flows. However, in TAFA, the data flow and the acknowledgment flow are regarded as the original flow and derivative flow, respectively, and special processing is invoked as discussed in Section 2.2.2.4. Simulation results are shown in Table 2.8 for only the configurations when the IEEE 802.11 MAC protocol has fairness problems.

It is clear from Table 2.8 that the fairness problem is much more severe for two competing TCP-based flows than for the case of UDP-based flows if no special

Table 2.8. Throughput comparison for the IEEE 802.11, the hybrid scheme and TAFA – two FTP flows (throughput measured in kbps)

Config #	Scheme	Flow #	Throughput	Flow #	Throughput	Aggregate
4-1	802.11	$0 \rightarrow 1$	0	$2 \rightarrow 3$	926	929
	Hybrid	$0 \rightarrow 1$	-	$2 \rightarrow 3$	-	-
	TAFA	$0 \rightarrow 1$	249	$2 \rightarrow 3$	423	672
4-2	802.11	$1 \rightarrow 0$	488 ± 103	$2 \rightarrow 3$	453 ± 102	942
	Hybrid	$0 \rightarrow 1$	439 ± 99	$3 \rightarrow 2$	502 ± 98	940
	TAFA	$0 \rightarrow 1$	383	$3 \rightarrow 2$	390	773
4-3	802.11	$0 \rightarrow 1$	530 ± 432	$3 \rightarrow 2$	392 ± 438	922
	Hybrid	$0 \rightarrow 1$	397 ± 71	$3 \rightarrow 2$	455 ± 78	852
	TAFA	$0 \rightarrow 1$	272	$3 \rightarrow 2$	363	635
4-7	802.11	$0 \rightarrow 1$	928	$2 \rightarrow 3$	0	930
	Hybrid	$0 \rightarrow 1$	-	$2 \rightarrow 3$	-	-
	TAFA	$0 \rightarrow 1$	443	$2 \rightarrow 3$	332	775
4-8	802.11	$0 \rightarrow 1$	929	$3 \rightarrow 2$	0	930
	Hybrid	$0 \rightarrow 1$	-	$3 \rightarrow 2$	-	-
	TAFA	$0 \rightarrow 1$	409	$3 \rightarrow 2$	209	617
4-10	802.11	$0 \rightarrow 1$	376	$3 \rightarrow 2$	526	902
	Hybrid	$0 \rightarrow 1$	-	$3 \rightarrow 2$	-	-
	TAFA	$0 \rightarrow 1$	335	$3 \rightarrow 2$	438	773

processing is in place. For example, in some cases, such as configurations 4-1, 4-7 and 4-8, one FTP flow is denied access to the shared channel for most of the time. The hybrid scheme, due to the lack of flow contention information, cannot trigger the desired receiver-initiated collision avoidance handshake, hence it is of no avail. On the other hand, TAFA achieves much better fairness though at a cost of degraded throughput. This is a much desired tradeoff because it avoids the starvation of some flows and hence channel bandwidth is more evenly distributed among participating nodes. For configuration 4-3, please note the high variation of the throughput for these two flows in the case of IEEE 802.11 which shows that one flow monopolizes the channel for a long time and then gives it away to the other flow. Both the hybrid scheme and TAFA help to solve the problem. For other configurations, TAFA suffers some degradation in throughput. However, the overall performance of TAFA shows a much better tradeoff between throughput and fairness among the three schemes we investigate. We expect that even better algorithms than TAFA can be designed in the future following our fairness framework.

2.3 Conclusion

In this chapter, we have presented our work on throughput and fairness of collision avoidance protocols in ad hoc networks.

In the first part of our work, we use a simple model to derive the saturation throughput of MAC protocols based on an RTS-CTS-data-ACK handshake in multi-hop networks. The results show that these protocols outperform CSMA protocols, even when the overhead of RTS/CTS exchange is rather high, thus showing the importance of correct collision avoidance in random access protocols. More importantly, it is shown that the overall performance of the sender-initiated collision avoidance scheme degrades rather rapidly when the number of competing nodes allowed within a region increases, in contrast to the case of fully-connected networks and networks with limited hidden terminals reported in the literature [11, 10, 13], where throughput remains almost the same for a large number of nodes. The significance of the analysis is that the scalability problem of contention-based collision-avoidance MAC protocols looms much earlier than people might expect. Simulation experiments with the IEEE 802.11 MAC protocol validate these observations and show that the IEEE 802.11 MAC protocol can suffer severe degradation in throughput due to its inability to avoid collisions between data packets and other packets even when the number of competing nodes in a region is small. However, when the number of competing nodes in a region increases, the performance gap is smaller as perfect collision avoidance protocols also begins to suffer from exceedingly long waiting time.

In the second part of our work, we propose a framework to address the fairness problem in ad hoc networks systematically. The framework includes four key components: Exchange of flow contention information, adaptive and stable backoff algorithm, hybrid collision avoidance handshake, and special processing for two-way flows. We proposed some specific algorithms to realize the framework and the resulting scheme, called topology aware fair access (TAFA), was evaluated through computer simulations against the IEEE 802.11 MAC protocol and a hybrid channel scheme proposed earlier in the literature. It was shown that TAFA can solve the fairness problem in UDP-based applications with negligible degradation in throughput. TAFA is also quite promising for TCP-based applications, which have not been investigated at length in the past. Though TAFA suffers some throughput degradation, it solves the notorious problem of starvation of TCP flows, thus showing a much better overall tradeoff between throughput and fairness than the other schemes.

Given that the fairness framework is tailored to ad hoc networks and is general enough to accommodate new algorithms, it will be interesting to investigate new adaptive backoff algorithm and new criteria to switch between sender-initiated and receiver-initiated collision avoidance to achieve better throughput and fairness tradeoffs in future work.

References

[1] V. Bharghavan, A. Demers, S. Shenker, and L. Zhang, "MACAW: A Media Access Protocol for Wireless LANs," in *Proc. of ACM SIGCOMM '94*, 1994.

[2] IEEE Computer Society LAN MAN Standards Committee, ed., *IEEE Standard for Wireless LAN Medium Access Control (MAC) and Physical Layer (PHY) Specifications.* IEEE Std 802.11-1997, The Institute of Electrical and Electronics Engineers, New York, 1997.

[3] J. J. Garcia-Luna-Aceves and C. L. Fullmer, "Floor Acquisition Multiple Access (FAMA) in Single-channel Wireless Networks," *ACM/Baltzer Mobile Networks and Applications*, vol. 4, no. 3, pp. 157–174, 1999.

[4] F. Talucci and M. Gerla, "MACA-BI (MACA by Invitation): A Receiver Oriented Access Protocol for Wireless Multihop Networks," in *Proc. of PIMRC '97*, 1997.

[5] J. J. Garcia-Luna-Aceves and A. Tzamaloukas, "Receiver-initiated Collision Avoidance in Wireless Networks," *ACM Wireless Networks*, vol. 8, pp. 249–263, 2002.

[6] F. A. Tobagi and L. Kleinrock, "Packet Switching in Radio Channels: Part II - the Hidden Terminal Problem in Carrier Sense Multiple-access Modes and the Busy-tone Solution," *IEEE Trans. on Communications*, vol. 23, no. 12, pp. 1417–1433, 1975.

[7] Y. Wang and J. J. Garcia-Luna-Aceves, "Performance of Collision Avoidance Protocols in Single-Channel Ad Hoc Networks," in *Proc. of IEEE Intl. Conf. on Network Protocols (ICNP '02)*, (Paris, France), Nov. 2002.

[8] H. Takagi and L. Kleinrock, "Optimal Transmission Range for Randomly Distributed Packet Radio Terminals," *IEEE Transactions on Communications*, vol. 32, no. 3, pp. 246–57, 1984.

[9] L. Wu and P. Varshney, "Performance Analysis of CSMA and BTMA Protocols in Multihop Networks (I). Single Channel Case," *Information Sciences, Elsevier Sciences Inc.*, vol. 120, pp. 159–77, 1999.

[10] F. Cali, M. Conti, and E. Gregori, "Dynamic Tuning of the IEEE 802.11 Protocol to Achieve a Theoretical Throughput Limit," *IEEE/ACM Transactions on Networking*, vol. 8, pp. 785–799, Dec. 2000.

[11] G. Bianchi, "Performance Analysis of the IEEE 802.11 Distributed Coordination Function," *IEEE Journal on Selected Areas in Communications*, vol. 18, pp. 535–547, Mar. 2000.

[12] X. Zeng, R. Bagrodia, and M. Gerla, "GloMoSim: a Library for Parallel Simulation of Large-scale Wireless Networks," in *Proc. of the 12th Workshop on Parallel and Distributed Simulations*, May 1998.

[13] F. Cali, M. Conti, and E. Gregori, "IEEE 802.11 Protocol: Design and Performance Evaluation of an Adaptive Backoff Mechanism," *IEEE Journal on Selected Areas in Communications*, vol. 18, pp. 1774–1786, Sept. 2000.

[14] T. Ozugur, M. Naghshineh, P. Kermani, C. M. Olsen, B. Rezvani, and J. A. Copeland, "Balanced Media Access Methods for Wireless Networks," in *Proc. of ACM/IEEE MOBICOM '98*, pp. 21–32, Oct. 1998.

[15] B. Bensaou, Y. Wang, and C. C. Ko, "Fair Medium Access in 802.11 Based Wireless Ad-Hoc Networks," in *IEEE/ACM Intl. Workshop on Mobile Ad Hoc Networking and Computing (MobiHoc '00)*, (Boston, MA, U.S.A.), Aug. 2000.

[16] X. Huang and B. Bensaou, "On Max-min Fairness and Scheduling in Wireless Ad-Hoc Networks: Analytical Framework and Implementation," in *ACM MobiHoc '01*, (Long Beach, CA, U.S.A.), Oct. 2001.

[17] N. H. Vaidya, P. Bahl, and S. Gupta, "Distributed Fair Scheduling in a Wireless LAN," in *ACM Mobicom 2000*, (Boston, MA, USA), Aug. 2000.

[18] T. Nandagopal, T. Kim, X. Gao, and V. Bharghavan, "Achieving MAC Layer Fairness in Wireless Packet Networks," in *ACM Mobicom 2000*, (Boston, MA, USA), Aug. 2000.

[19] H. Luo, S. Lu, and V. Bharghavan, "A New Model for Packet Scheduling in Multihop Wireless Networks," in *ACM Mobicom 2000*, (Boston, MA, USA), Aug. 2000.

[20] H. Luo and S. Lu, "A Topology-Independent Fair Queueing Model in Ad Hoc Wireless Networks," in *IEEE ICNP 2000*, (Osaka, Japan), Nov. 2000.

[21] H. Luo, P. Medvedev, J. Cheng, and S. Lu, "A Self-Coordinating Approach to Distributed Fair Queueing in Ad Hoc Wireless Networks," in *IEEE INFOCOM 2001*, Apr. 2001.

[22] Y. Wang and J. J. Garcia-Luna-Aceves, "A New Hybrid Channel Access Scheme for Ad Hoc Networks," *ACM Wireless Networks Journal, Special Issue on Ad Hoc Networking*, vol. 10, no. 4, 2004.

Chapter 3

ROUTING IN MOBILE AD HOC NETWORKS

Mahesh K. Marina and Samir R. Das
Department of Computer Science
State University of New York at Stony Brook
Stony Brook, NY 11794-4400

Abstract

Efficient, dynamic routing is one of the key challenges in mobile ad hoc networks. In the recent past, this problem was addressed by many research efforts, resulting in a large body of literature. We survey various proposed approaches for routing in mobile ad hoc networks such as flooding, proactive, on-demand and geographic routing, and review representative protocols from each of these categories. We further conduct qualitative comparisons across various approaches. We also point out future research issues in the context of individual routing approaches as well as from the overall system perspective.

Keywords: Unicast routing, mobile ad hoc networks, multihop wireless networks, packet radio networks, flooding, proactive routing, on-demand routing, geographic routing.

3.1 Introduction

Developing support for routing is one of the most significant challenge in ad hoc networks and is critical for the basic network operations. Certain unique combinations of characteristics make routing in ad hoc networks interesting. First, nodes in an ad hoc network are allowed to move in an uncontrolled manner. Such node mobility results in a highly dynamic network with rapid topological changes causing frequent route failures. A good routing protocol for this network environment has to dynamically adapt to the changing network topology. Second, the underlying wireless channel provides much lower and more variable bandwidth than wired networks. The wireless channel working as a shared medium makes available bandwidth per node even lower. So routing protocols should be bandwidth-efficient by expending a minimal overhead for computing routes so that much of the remaining bandwidth is available for the actual data communication. Third, nodes run on batteries which have limited energy supply. In order for nodes to stay and communicate for longer periods, it is desirable that a routing protocol be energy-efficient as well. This also provides also another reason why overheads must be kept low. Thus, routing protocols must meet the conflicting goals of dynamic adaptation and low overhead to deliver good overall performance.

Routing protocols developed for wired networks such as the wired Internet are inadequate here as they not only assume mostly fixed topology but also have high overheads[1]. This has lead to several routing proposals specifically targeted for ad hoc networks. While some of these proposals are optimized variants of protocols originally designed for wired networks, the rest adopt new paradigms such as on-demand routing, where routes are maintained "reactively" only when needed. This is in contrast with the traditional, proactive Internet-based protocols. Other new paradigms also have emerged – for example, exploiting location information fro routing, and energy-efficient routing.

All our discussions here implicitly assume that underlying network topology can be viewed as an undirected graph. In practice, this assumption may not hold, since unidirectional links may be present. This commonly occurs when there is a difference in the transmit powers in the nodes of the network. Even in a perfectly homogeneous network, interference at the wireless channel can be spatially diverse, causing unidirectionality of links. However, there is both empirical [38] and theoretical [3] evidence showing that using unidirectional links for routing may not yield any substantial benefit. On the contrary, using such links is complex and may increase overheads. On the other hand, ignoring unidirectional links, when indeed present, is straightforward. It can be realized via simple two-way message exchanges between neighboring nodes. Many routing protocols ordinarily exchange such messages (often called "beacons" or "hello" messages) for the purpose of finding the neighbor node set (neighbor discovery).

Routing research in ad hoc networks is quite broad. In this chapter, we limit ourselves to unicast routing and associated techniques, and do not discuss multi-destination routing such as multicast or geocast. Fundamental routing issues can be understood quite well by developing a good background on unicast routing issues and techniques. The rest of this chapter is organized as follows. In the next Section, we discuss flooding and a few efficient variants of basic flooding — flooding is not only a legitimate candidate for unicast routing in extremely mobile scenarios, but also is an integral part of several other routing protocols. In Section 3, we will review optimized variants of traditional distance

[1]Routing protocols for satellite networks are also inadequate here as the topology in satellite networks is completely deterministic at any time even though nodes are moving.

vector and link state protocols tailored for ad hoc networks. Section 4 describes three prominent on-demand protocols along with some generic optimizations. In Section 5, we compare proactive and reactive approaches and also discuss hybrid approaches that attempt to combine the benefits of these two approaches. In Section 6, we discuss routing using geographic location information. We finally conclude in Section 7.

3.2 Flooding

Flooding (or network-wide broadcasting) is the simplest way to deliver data from a node to any other node in the network. In flooding, the source simply broadcasts the data packet to its neighboring nodes via a MAC layer broadcast mechanism. Each node hearing the broadcast *for the first time* re-broadcasts it. Thus, the broadcast propagates in "layers" outwards from the source, eventually terminating when every node has heard the packet and transmitted it once. The rule "every node transmit only once" guarantees termination of the procedure and also avoids looping. This can be achieved using unique identifiers on all packets being flooded. The flooding technique delivers the data to every node in the connected component of the network.

With flooding, no topological information needs to be maintained or known in advance. In network scenarios where node mobility is so high that a given unicast routing protocol may fail to keep up with the rate of topology changes, flooding may become the only alternative for routing data reasonably. However, in other scenarios where node mobility is trackable by a routing protocol, flooding can be a very inefficient option. This is because the total number of transmissions to deliver a single message to a destination with flooding is in the order of network size, as opposed to the network diameter with a unicast routing protocol (assuming that a route is already found).

Although flooding is not usually attractive for efficiently delivering data, it is still very useful in carrying out certain routing tasks such as route discovery and topology dissemination, and as a bootstrapping mechanism when nothing is known a priori about the network topology. Therefore, flooding appears as a key component in many routing protocols (OSPF [40] is a classic example).

In the simple flooding protocol as described above (also called *pure flooding*), each node transmits (broadcasts) the data once. As a result, a node may receive the same packet from several neighbors. Thus, depending on the network density, simple flooding may take far more transmissions than necessary for the flood to reach every node. Such redundancy can be eliminated to achieve less contention and collisions at the radio link layer, thus increasing network utilization. Several efficient alternatives have been proposed that use only a small subset of nodes to transmit the data packet during a flood, however ensure that all nodes in the network receive the packet.

3.2.1 Efficient Flooding Techniques

Efficient flooding essentially attempts to eliminate the redundant broadcasts, but still ensures that all nodes in the network receives the packet. In the simplest of all techniques, every node (other than the source) rebroadcasts the data packet only with a certain probability p [42] [17]. Clearly, the correct choice of the probability p determines the effectiveness of this technique — a very small value prevents the flood from reaching every node ("flood dying out" problem), while a very large value results in many redundant broadcasts. The right value of p depends on several factors, including the average node degree. Haas et. al. [17] evaluate several variants of this basic technique and show that with appropriate choice of p, that changes as the flood propagates away from the source, significant savings are possible without affecting the coverage of the flood (i.e., number of nodes receiving the flooded packet). Ideally, the flood should cover all nodes in the network. Determining the right value of p remains a hard problem.

Other techniques are also possible. For example, when a node hears the broadcast packet, it does not transmit it immediately, but waits for a brief period to see whether it hears the same packet again. If it does hear it multiple times (say, k times) within this period, it assumes that all its neighbors must have heard this packet, and refrains from transmitting it [42]. As before, determining suitable values for k and the waiting period becomes complex. This technique can be improved by incorporating neighborhood knowledge. For example, if each node knows its neighbor set and this set is included in each broadcast, then it is easy for a node to completely determine whether all its neighbors have heard this packet by computing the union of the neighbor sets transmitted in the packets it has heard. Still, how long a node should wait to hear all the broadcasts from its neighbors remains a question.

This problem of eliminating redundant broadcasts can be solved via a more algorithmic approach. The objective is to determine a small subset of nodes for broadcasting data such that every node in the network receives it. Often this subset is called the *forwarding set*. This problem is equivalent to finding a *Minimum Connected Dominating Set (MCDS)*[2]. In a Dominating Set (DS), a node is either designated as a dominator or is a neighbor of at least one dominator node. A Connected Dominating Set (CDS) is a DS such that the subgraph formed by considering only the dominator nodes (and edges among them) is connected. MCDS is a CDS of the smallest size. Even with full topology information, the MCDS problem is difficult and shown to be NP-hard [14]. Therefore, research efforts have focused on developing efficient

[2]This problem is also closely related to maximum leaf spanning tree problem and a special case of minimum set cover problem

centralized approximation algorithms [16] and distributed heuristics using only partial and local topology information [32] [54] [51] [46] [59].

Distributed heuristics for efficient flooding (also termed *neighborhood-knowledge techniques*) seek to find a small forwarding set without incurring too much overhead in the process. Some heuristics explicitly find a CDS [32] [54], while others do it implicitly [51] [46]. The implicit heuristics can be further classified into two categories: neighbor designating methods (e.g., [51]) and self-pruning (e.g., [46]). In neighbor designating nodes, the status of whether a node should be in the forward set is determined by its neighbors. On the other hand, each node determines its own status in self-pruning methods. These heuristics also differ in the time the forwarding set is computed. Some heuristics dynamically compute the forward node set (e.g., [46]) depending on the source of the flood and neighboring nodes that have already rebroadcasted, while other heuristics determine the forward node set statically (e.g., [51]) independent to any specific source.

Williams and Camp [58] have compared the performance of several efficient flooding techniques including the probabilistic flooding technique described earlier. They found that neighborhood-knowledge techniques in general perform better than probabilistic technique, especially in low and moderate mobility scenarios. The effectiveness of the neighborhood-knowledge methods reduces at high mobility because of inaccuracy in neighborhood information used in finding the forward node set. In fact, some amount of controlled redundancy in the forward node set is beneficial in coping with mobility and unreliability of broadcasts in some MAC protocols such as IEEE 802.11 DCF [24]. Recently, there is also some work on unifying different neighborhood-knowledge techniques into a single generic broadcast scheme by recognizing that they share similar features [59]. We will discuss just one scheme, called *Multipoint Relaying*, in some detail here to give a flavor of the available techniques. We have chosen this particular scheme since it is the key component of a routing protocol to be discussed later.

In Multipoint Relaying [51], the main idea is that each node selects a small subset of its neighbors as Multipoint Relays (MPRs) sufficient to cover its 2-hop neighborhood (Figure 3.1). When a node floods a packet, only the MPRs of the node rebroadcast the packet and their MPRs rebroadcast and so on. Nodes exchanges their list of neighbors via periodic "hello" packets. As a result, each node knows its 2-hop neighborhood information. Each node then locally computes its MPR set using the following heuristic because finding the minimum size MPR set is NP-hard. The node includes a neighbor in its MPR set if it is the only neighbor to reach a 2-hop neighbor. After including all such neighbors, the node picks a neighbor not already in the MPR set which can cover the most number of nodes that are uncovered by the current MPR set. The node repeats this last step until all 2-hop neighbors are covered by

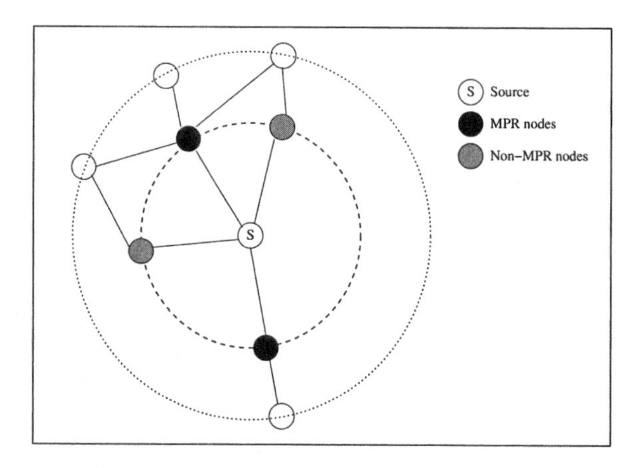

Figure 3.1. Multipoint Relay concept. Two dotted circles around the source S represent its logical 1-hop and 2-hop neighborhood respectively.

the MPR set. The node then informs the neighbors selected as MPRs via hello packets and it becomes the MPR-Selector for those neighbors. Every node is also responsible for updating its MPR set and notifying the corresponding neighbors whenever the neighborhood changes. It is also shown that MPR set computed by the above heuristic is within a $\log(n)$ bound of the optimal size set, where n is the network size.

3.3 Proactive Routing

Proactive protocols maintain unicast routes between all pairs of nodes regardless of whether all routes are actually used. Therefore, when the need arises (i.e., when a traffic source begins a session with a remote destination), the traffic source has a route readily available and does not have to incur any delay for route discovery. These protocols also can find optimal routes (shortest paths) given a model of link costs.

Routing protocols on the Internet (i.e, distance vector-based RIP [19] and link state-based OSPF [40]) fall under this category. However, these protocols are not directly suitable for resource-poor and mobile ad hoc networks because of their high overheads and/or somewhat poor convergence behavior. Therefore, several optimized variations of these protocols have been proposed for use in ad hoc networks. These protocols are broadly classified into the two traditional categories: distance vector and link state. In distance vector protocols, a node exchanges with its neighbors a vector containing the current distance information to all known destinations; the distance information propagates across the network transitively and routes are computed in a distributed manner at each node. On the other hand, in link state protocols, each node disseminates the status of each of its outgoing links throughout the network (typically via flooding) in the form of link state updates. Each node locally computes routes in a decentralized manner using the complete topology information. In the rest of this section, we describe two protocols from each of these categories that have received wide attention.

3.3.1 Distance Vector Protocols

Destination-Sequenced Distance-Vector (DSDV) [48] was one of the earliest protocols developed for ad hoc networks. Primarily design goal of DSDV was to develop a protocol that preserves the simplicity of RIP, while guaranteeing loop freedom. It is well known that Distributed Bellman-Ford (DBF) [2], the basic distance vector protocol, suffers from both short-term and long-term routing loops (the *counting-to-infinity* problem) and thus exhibits poor convergence in the presence of link failures. Note that RIP is DBF with the addition of two ad hoc techniques (split-horizon and poisoned-reverse) to prevent two hop loops. The variants of DBF proposed to prevent loops (Merlin-Segall [39], Jaffe-Moss [25], and DUAL [13]), however, involve complex inter-nodal coordination. Because of inter-nodal coordination, the overheads of these proposals are much higher than basic DBF and match that of link-state protocols using flooding to disseminate link-state updates; so, these protocols are effective only when topology changes are rare.

The main idea in DSDV is the use of destination sequence numbers to achieve loop freedom without any inter-nodal coordination. Every node maintains a monotonically increasing sequence number for itself. It also maintains the highest known sequence number for each destination in the routing table (called "destination sequence numbers"). The distance/metric information for every destination, typically exchanged via routing updates among neighbors in distance-vector protocols, is tagged with the corresponding destination sequence number. These sequence numbers are used to determine the relative freshness of distance information generated by two nodes for the same destination (the node with a higher destination sequence number has the more recent information). Routing loops are prevented by maintaining an invariant that destination sequence numbers along any valid route monotonically increase toward the destination.

DSDV also uses triggered incremental routing updates between periodic full updates to quickly propagate information about route changes. In DSDV, like in DBF, a node may receive a route with a longer hop count earlier than the one with the smallest hop count. Therefore, always propagating distance information immediately upon change can trigger many updates that will ripple through the network, resulting in a huge overhead. So, DSDV estimates route settling time (time it takes to get the route with the shortest distance after getting the route with a higher distance) based on past history and uses it to avoid propagating every improvement in distance information.

Wireless Routing Protocol (WRP) [41] is another distance vector protocol optimized for ad hoc networks. WRP belongs to a class of distance vector protocols called *path finding algorithms*. The algorithms of this class use the next hop and second-to-last hop information to overcome the counting-to-infinity problem; this information is sufficient to locally determine the shortest path spanning tree at each node. In these algorithms, every node is updated with the shortest path spanning tree of each of its neighbors. Each node uses the cost of its adjacent links along with shortest path trees reported by neighbors to update its own shortest path tree; the node reports changes to its own shortest path tree to all the neighbors in the form of updates containing distance and second-to-last hop information to each destination.

Path finding algorithms originally proposed for the Internet (e.g., [8]) suffer from temporary routing loops even though they prevent the counting-to-infinity problem. This happens because these algorithms fail to recognize that updates received from different neighbors may not agree on the second-to-last hop to a destination. WRP improves on the earlier algorithms by verifying the consistency of second-to-last hop reported by all neighbors. With this mechanism, WRP reduces the possibility of temporary routing loops, which in turn results in faster convergence time. One major drawback of WRP is its requirement for reliable and ordered delivery of routing messages.

3.3.2 Link State Protocols

Optimized Link State Routing (OLSR) [9] is an optimized version of traditional link state protocol such as OSPF. It uses the concept of Multipoint Relays (MPRs), discussed in the previous section, to efficiently disseminate link state updates across the network. Only the nodes selected as MPRs by some node are allowed to generate link state updates. Moreover, link state updates contain only the links between MPR nodes and their MPR-Selectors in order to keep the update size small. Thus, only partial topology information is made available at each node. However, this information is sufficient for each to locally compute shortest hop path to every other node because at least one such path consists of only MPR nodes.

OLSR uses only periodic updates for link state dissemination. Since the total overhead is then determined by the product of number of nodes generating the updates, number of nodes forwarding each update and the size of each update, OLSR reduces the overhead compared to a base link state protocol when the network is dense. For a sparse network, OLSR degenerates to traditional link state protocol. Finally, using only periodic updates makes the choice of update interval critical in reacting to topology changes.

Topology Broadcast based on Reverse-Path Forwarding (TBRPF) [43] is a partial topology link state protocol where each node has only partial view of the whole network topology, but sufficient to compute a shortest path source spanning tree rooted at the node. When a node obtains source trees maintained at neighboring nodes, it can update its own shortest path tree. This idea is somewhat similar to that in path finding algorithms such as WRP discussed above. TBRPF exploits an additional fact that shortest path trees reported by neighbors can have a large overlap. A node can still compute its shortest path tree even if it receives partial trees from each of its neighbors as long as they minimally overlap. Thus, every node reports only a part of its source tree (called Reported Tree (RT)) to all neighbors in an attempt to reduce the size of topology updates. A node uses periodic topology updates to inform its complete RT to all neighbors at longer intervals, while it uses differential updates to inform them about the changes to its RT more frequently.

In order to compute RT, a node X first determines a Reported Node (RN) set. RN contains itself (node X) and each neighbor Y for which X is on the shortest path to Y from another neighbor. RN so computed contains X and a subset (possibly empty) of its neighbors. For each neighbor Y included in RN, X acts as a forwarding node for data destined to Y. Finally, X also includes in RN all nodes which can be reached by a shortest path via one of its neighbors already in RN. Once X completes computing RN as stated above, the set of all links (u, v) such that $u \in$ RN, constitute the RT of X. Note that RT only specifies the minimum amount of topology that a node must report to its neighbors. To obtain some redundancy in the topology maintained at each node (e.g., a subgraph more connected than a tree), nodes can report more topology than RT.

TBRPF also employs an efficient neighbor discovery mechanism using differential hellos for nodes to determine their bidirectional neighbors. This mechanism reduces the size of hello messages by avoiding the need to include every neighbor in each hello message.

3.3.3 Performance of Proactive Protocols

Among the proactive protocols we have discussed, DSDV seems to suffer from poor responsiveness to topology changes and slow convergence to optimal paths. This is mainly because of the transitive nature of topology updates in distance vector protocols. Simulation results [5] [26] also confirm this behavior. Although reducing the update intervals appears to improve its responsiveness, it might also proportionately increase the overhead leading to congestion. WRP, the other distance vector protocol we have discussed, assumes reliable and in-order delivery of routing control packets which is an unreasonable requirement in error-prone wireless networks. The performance of the protocol when this assumption does not hold is unclear. As far as the two link state protocols

— OLSR and TBRPF — are concerned, both of them share some features such as being partial topology protocols. However, the details of the protocols are quite different. Whereas OLSR is more like a traditional link state protocol with optimizations to reduce overhead in ad hoc networks, TBRPF is a link state variant based on tree sharing concept. TBRPF also has one desirable feature of using frequent incremental updates in addition to periodic, less frequent full updates. This feature will likely improve responsiveness to topology changes. OLSR, on the other hand, relies solely on periodic full updates. Although in our knowledge there is no comprehensive study focusing on relative performance of OLSR and TBRPF, they expected to show comparable performance (and likely better than their distance vector counterparts).

3.4 On-demand Routing

On-demand (reactive) routing presents an interesting and significant departure from the traditional proactive approach. Main idea in on-demand routing is to find and maintain *only needed* routes. Recall that proactive routing protocols maintain all routes without regard to their ultimate use. The obvious advantage with discovering routes on-demand is to avoid incurring the cost of maintaining routes that are not used. This approach is attractive when the network traffic is sporadic, bursty and directed mostly toward a small subset of nodes. However, since routes are created when the need arises, data packets experience queuing delays at the source while the route is being found at session initiation and when route is being repaired later on after a failure. Another, not so obvious consequence of on-demand routing is that routes may become suboptimal, as time progresses since with a pure on-demand protocol a route is used until it fails. In the rest of this Section, we describe three well-known on-demand protocols and follow them up with some generic set of optimizations that can benefit any on-demand protocol.

3.4.1 Protocols for On-Demand Routing

Dynamic Source Routing (DSR) [27] [28] is characterized by the use of *source routing*. That is, the sender knows the complete hop-by-hop route to the destination. These routes are stored in a *route cache*. The data packets carry the source route in the packet header.

When a node in the ad hoc network attempts to send a data packet to a destination for which it does not already know the route, it uses a *route discovery* process to dynamically determine such a route. Route discovery works by *flooding* the network with *route request* (also called *query*) packets. Each node receiving a request, rebroadcasts it, unless it is the destination or it has a route to the destination in its route cache. Such a node replies to the request with a *route reply* packet that is routed back to the original source. Route request and reply packets are also source routed. The request builds up the path traversed so far. The reply routes itself back to the source by traversing this path backward. The route carried back by the reply packet is cached at the source for future use. If any link on a source route is broken (detected by the failure of an attempted data transmission over a link, for example), a *route error* packet is generated. Route error is sent back toward the source which erases all entries in the route caches along the path that contains the broken link. A new route discovery must be initiated by the source, if this route is still needed and no alternate route is found in the cache.

DSR makes aggressive use of source routing and route caching. With source routing, complete path information is available and routing loops can be easily detected and eliminated without requiring any special mechanism. Because route requests and replies are both source routed, the source and destination, in addition to learning routes to each other, can also learn and cache routes to all intermediate nodes. Also, any forwarding node caches any source route in a packet it forwards for possible future use. DSR employs several optimizations including promiscuous listening which allows nodes that are not participating in forwarding to overhear on-going data transmissions nearby to learn different routes free of cost. To take full advantage of route caching, DSR replies to *all* requests reaching a destination from a single request cycle. Thus the source learns many alternate routes to the destination, which will be useful in the case that the primary (shortest) route fails. Having access to many alternate routes saves route discovery floods, which is often a performance bottleneck. This may, however, result in route reply flood unless care is taken.

However, aggressive use of route caching comes with a penalty. Basic DSR protocol lacks effective mechanisms to purge stale routes. Use of stale routes not only wastes precious network bandwidth for packets that are eventually dropped, but also causes cache pollution at other nodes when they forward/overhear stale routes. Several performance studies [20] [50] have shown that stale caches can significantly hurt performance especially at high mobility and/or high loads. These results have motivated subsequent work on improved caching strategies for DSR [21] [37] [23]. Besides stale cache problems, the use of source routes in data packets increases the byte overhead of DSR. This limitation was addressed in a later work by the DSR designers [22].

Ad hoc On-demand Distance Vector (AODV) [49] [47] shares DSR's on-demand characteristics in that it also discovers routes on an "*as needed*" basis via a similar route discovery process. However, AODV adopts a very different mechanism to maintain routing information. It uses traditional routing tables, one entry per destination. This is in contrast to DSR, which can maintain multiple route cache entries for each destination. Without source routing, AODV relies on routing table entries to propagate a RREP back to the source and, subsequently, to route data packets to the destination. AODV uses destination sequence numbers as in DSDV [48] (Section 3) to prevent routing loops and to determine freshness of routing information. These sequence numbers are carried by all routing packets.

The absence of source routing and promiscuous listening allows AODV to gather only a very limited amount of routing information with each route discovery. Besides, AODV is conservative in dealing with stale routes. It uses the sequence numbers to infer the freshness of routing information and nodes maintain only the route information for a destination corresponding to the latest known sequence number; routes with older sequence numbers are discarded even though they may still be valid. AODV also uses a timer-based route expiry mechanism to promptly purge stale routes. Again if a low value is chosen for the timeout, valid routes may be needlessly discarded.

In AODV, each node maintains at most one route per destination and as a result, the destination replies only once to the first arriving request during a route discovery. Being a single path protocol, it has to invoke a new route discovery whenever the only path from the source to the destination fails. When topology changes frequently, route discovery needs to be initiated often which can be very inefficient since route discovery flood is associated with significant latency and overhead. To overcome this limitation, we have proposed a multipath extension to AODV called Ad hoc On-demand Multipath Distance Vector (AOMDV) [36]. AOMDV discovers multiple paths between source and destination in a single route discovery. As a result, a new route discovery is necessary only when each of the multiple paths fail. AOMDV, like AODV, ensures loop freedom and at the same time finds disjoint paths which are less likely to fail simultaneously. By exploiting already available alternate path routing information as much as possible, AOMDV computes alternate paths with minimal additional overhead over AODV.

Temporally Ordered Routing Algorithm (TORA) [44] is another on-demand protocol. TORA's route discovery procedure computes multiple loop-free routes to the destination which constitute a *destination-oriented directed acyclic graph* (DAG).

While the ad hoc network is looked upon as an undirected graph, TORA imposes a logical directionality on the links. TORA employs a route maintenance procedure requiring strong inter-nodal coordination based on a *link reversal* concept proposed in a seminal work by Gafni and Bertsekas [12] for localized recovery from route failures. The basic idea behind link reversal algorithms is as follows. Whenever a link failure at a node causes the node to lose all downstream links to reach the destination (and thus no longer in a destination-oriented state), a series of link reversals starting at that node can revert the DAG back to a destination-oriented state.

There are two types of link reversal algorithms namely full reversal and partial reversal differing in the way links incident on a node reverse their direction during the link reversal process. TORA specifically uses a modified version of partial link reversal technique. This modified version allows TORA to detect network partitions, a useful feature absent in many ad hoc networking protocols.

By virtue of finding multiple paths and using the link reversals for recovering from route failures, TORA can avoid a fresh route discovery until all paths connecting the source and the destination break (which is similar to AOMDV [36]). But TORA requires reliable and in-order delivery of routing control packets. Also, the nature of link reversal based algorithm makes it hard to keep track of path costs. Some performance studies [5] [10] have shown that these requirements hurt the performance of TORA so much so that they undermine the advantage of having multiple paths. Also, the link reversal in TORA by its nature leads to short-term routing loops. However, TORA remains an attractive option when a large number of nodes must maintain paths directed to a chosen destination.

3.4.2 Optimizations for On-demand Routing

Several general purpose optimizations have been proposed for on-demand routing that are largely independent of any specific protocol. These optimizations can be classified into three categories: flooding optimizations, stable route selection, and route maintenance optimizations. We will give a brief overview of techniques in each of these categories below.

In describing various protocols in the previous section, we have assumed that simple flooding is used for route discovery. However, efficient flooding techniques discussed in Section 2 can be used to reduce route discovery overhead. But when neighborhood-knowledge techniques are employed, the overall benefit depends on the relationship between the frequency of route discovery operations, network density, and the overhead incurred in maintaining up-to-date neighborhood information at each node.

Recognizing that route discovery flood is intended to search only the destination offers more room for optimization since flood need not reach every node. *Expanding ring search* [47] and *query localization* [7] are two representative examples which exploit this fact by performing a restricted flood within a small region (relative to network size) containing both source and destination. In expanding ring search, source estimates the distance (in hops) to the destination and uses this estimated distance (ring size) in the form of TTL to do a limited flood around the source; when the route search fails, ring size is increased iteratively until the whole network is searched or a route is found. Simplest mechanism for distance estimation is to use the last known hop count to destination; more sophisticated procedures have also been studied [56]. Query localization, on the other hand, is based on the notion of spatial and temporal locality of paths. It makes the assumption that new path and broken old path

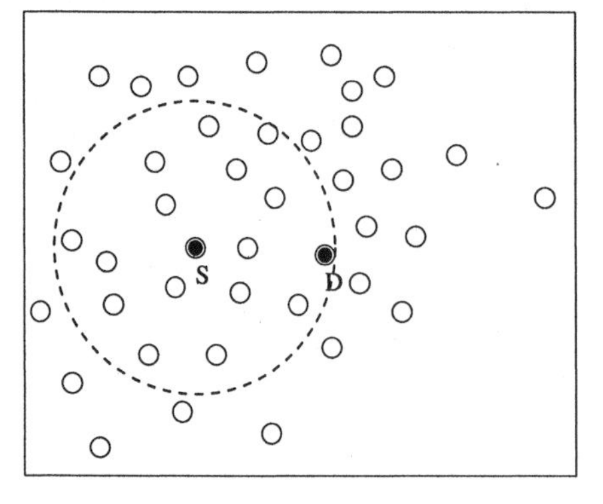

(a) Expanding ring search

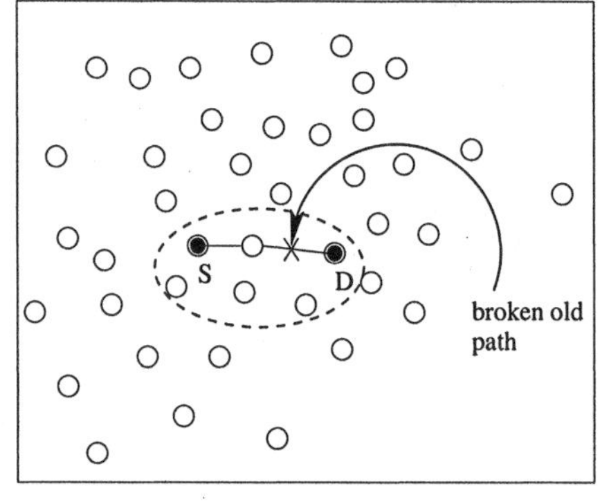

(b) Query localization

Figure 3.2. Comparison of search regions using expanding ring search and query localization. Dotted circles in each figure indicate the search regions.

will only differ in a few nodes and therefore, the query is restricted within a few hops around the old path; when route discovery fails, the search region is expanded in subsequent tries. Figure 3.2 illustrates the difference in search regions used by these two techniques.

All three protocols described in the previous section use hop counts as a metric for path selection. However, it is possible that the quality of the links on a shortest hop path is not be strong. The likelihood for this is not negligible because two neighboring nodes in a shortest hop path can be separated by physical distance almost equal to their transmission range. This not only makes the signal strength on the link weak, but also increases likelihood of path failure when either of them moves slightly away from the other node. This observation led to the work on better metrics for path selection. *Associativity-based Routing (ABR)* [57] and *Signal Stability-Based Adaptive Routing (SSR)* [11] are among the earliest protocols with the goal of long-lived route selection. ABR uses a link metric called degree of association stability which is calculated as the number of successful beacon exchanges between neighbors sharing a link in some interval; more beacon exchanges indicate a stable link and such links are preferred during route selection. In contrast, SSR uses signal strength information to determine link stability. In general, alternative metrics other than hop counts to determine path costs are possible. Suitable choice of metrics can serve other purposes, such as balancing load or energy usage in the network.

Local route repair (e.g., [47]) and *preemptive routing* [15] are key examples of route maintenance optimizations proposed for on-demand routing. In the local repair mechanism, the basic idea is to have an intermediate node repair a broken route locally. Using local repair, an intermediate node can find an alternate (possibly longer) route quickly and efficiently as compared to the source performing a new route discovery. The effectiveness of local repair depends on how far away the destination is from the intermediate node. Local repair can be done either reactively after a route failure or proactively. Preemptive routing, on the other hand, proactively repairs routes by monitoring the likelihood of a path break by means of signal strength information and informing the source which will initiate an early route discovery. Using this mechanism, applications will not experience the latency involved in discovering a route after the route breaks.

3.4.3 Performance of On-Demand Routing

Performance of on-demand protocols is quite well-understood. Some empirical performance results in literature have found TORA to be the worst performer among the three protocols we have discussed [5] [10]. TORA's link reversal technique, though elegant, requires strong inter-nodal coordination and thus has very high overhead. Besides, reliable and in-order delivery requirement imposes even greater demand in terms of bandwidth. As a result, later performance studies focused solely on the relative performance of DSR and AODV [50]. According to these studies, DSR with the help of caching is more effective at low mobility and low loads. AODV performs well in more stressful scenarios of high mobility and high loads. These relative performance differentials are attributed to DSR's lack of effective mechanisms to purge stale routes and AODV's need for resorting to route discovery often because of its single path nature. However, DSR with improved caching strategies, and AODV with the ability to maintain multiple paths are expected to have similar performance.

3.5 Proactive Versus On-demand Debate

As research on routing for ad hoc networks have matured, the superiority of one approach over the other has been debated. This question has motivated several simulation-based performance comparison studies [5] [10] [26] [4] and some theoretical studies (e.g., [52]). No clear winner emerged, although on-demand approach usually provides better or similar efficiency relatively for most common scenarios. Here we qualitatively compare the relative merits of the two approaches independently of any specific protocol.

Aggregate throughput and end-to-end delay are key measures of interest when assessing protocol performance. Throughput is directly related to the packet drops. Packet drops typically happen because of network congestion (e.g., buffer overflows) or for lack of a route. Since most dynamic protocols (proactive or reactive) try to keep the latter type (no route) of drops low by being responsive to topology changes, network congestion drops become the dominant factor when judging relative throughput performance. For the same data traffic load, routing protocol efficiency (in terms of control overhead in bytes or packets) determines the relative level of network congestion because both routing control packets and data packets share the same channel bandwidth and buffers.

End-to-end delay of a packet depends on route discovery latency, additional delays at each hop (comprising of queuing, channel access and transmission delays), and the number of hops. At low loads, queuing and channel access delays do not contribute much to the overall delay. In this regime, proactive protocols, by virtue of finding optimal routes between all node pairs, are likely to have better delay performance. However, at moderate to high loads, queuing and channel access delays become significant enough to exceed route discovery latency. So, like in the case of throughput, routing protocol overhead again becomes key factor in determining relative delay performance.

The efficiency (in terms of control overhead) of one approach over the other depends to a large extent on the relative node mobility and traffic diversity. Note that individual node speeds are irrelevant unless they affect the relative node speeds because path stability is primarily determined by relative node mobility; relative node mobility can be low even when nodes individually move at high speeds as with group movement scenarios. Traffic diversity measures the traffic distribution among nodes. A low traffic diversity indicates that majority of the traffic is directed toward a small subset of nodes. This can happen when there are fewer source-destination pairs communicating or when most of the nodes communicate with a few set of nodes. A realistic example of the latter case is when an ad hoc network is attached to the Internet and mobile nodes spend most time accessing the Internet via a few gateway nodes. High traffic diversity, on the other hand, means that traffic is more uniformly distributed across all nodes (e.g., when every node communicates with every other node).

On-demand routing is naturally adaptive to traffic diversity and therefore its overhead proportionately increases with increase in traffic diversity. On the other hand, for proactive routing overhead is independent of the traffic diversity. So when the traffic diversity is low, on-demand routing is relatively very efficient in terms of the control overhead regardless of relative node mobility. When the majority of traffic is destined to only few nodes, a proactive protocol maintaining routes to every possible destination incurs a lot of unnecessary overhead. Mobility does not alter this advantage of on-demand routing. This is because an on-demand protocol reacts only to link failures that break a currently used path, whereas proactive protocol reacts to every link failure without regard to whether the link is on a used path. On-demand routing can also significantly benefit by caching multiple paths when node mobility is low.

With high traffic diversity, the routing overhead for on-demand routing could approach that of proactive routing. The overhead alone is not the whole picture.

Path optimality also plays a role in determining the overall overhead — using a suboptimal path results in excess transmissions which contribute to overhead. Using suboptimal routes also increases the end-to-end delay. Recall that pure proactive protocols aim to always provide shortest paths. Whereas with pure on-demand protocols, a path is used until it becomes invalid even though the path may become suboptimal due to node mobility. The issue of path sub-optimality becomes more significant at low node mobility because each path is usable for a longer period. Thus, accounting suboptimal path overhead increases the total overhead with on-demand approach.

The discussion so far implicitly assumed that traffic sessions are long-lived. However, when traffic sessions are short-lived, i.e., come and go quickly, the overhead required to handle each session becomes expensive with on-demand routing. Also, initial route discovery latency inherent to on-demand routing may also be unacceptable in this case.

3.5.1 Hybrid Approaches

It is not hard to hypothesize that a combination of proactive and on-demand approaches is perhaps better than either approach in isolation. As an example, consider augmenting a primarily reactive protocol such as AODV with some proactive functionality by making each active destination periodically refresh routes to itself as in DSDV. The advantage of such a protocol is two-fold: (i) the overhead and delay due to suboptimal routes can be limited to the refresh interval; (ii) such destination-initiated refresh mechanism also offers routes in advance to nodes that might route traffic to the destination later, making this mechanism proactive. The overhead created by this proactive mechanism is determined by the number of active destinations and the refresh interval. Carefully choosing the refresh interval can improve the overall performance compared to the pure reactive mechanism. A variant of the above idea has been suggested in [44] and evaluated in [31]. There have been several other efforts based on this theme of combining proactive and reactive approaches. Below, we will review two representative protocols from this category. These two protocols, though mainly aim toward scalable routing for large networks, still demonstrate the benefit attainable by the combined proactive/reactive approach.

Zone Routing Protocol (ZRP) [18] [45] is a *hybrid* protocol with distinct proactive and reactive components working in cohesion. ZRP defines a zone for each node X which includes all nodes that are within a certain distance in hops, called zone radius, around the node X. Nodes that are exactly zone radius distance away from node X are called border nodes of X's zone. A proactive link state protocol is used to keep every node aware of the complete topology within its zone. When a node X needs to obtain a route to another node Y not in its zone, it reactively initiates a route discovery which works similar to flooding except that it involves only X's border nodes and their border nodes and so on. Route query accumulates the traversed route on its way outward from X (like in source routing) and when the query finally reaches a border node which is in destination Y's zone, that border node sends back a reply using the accumulated route from the query. Depending on the choice of zone radius, ZRP can behave as a pure proactive protocol, a pure reactive protocol, or somewhere in between. While this is an attractive feature to adapt to network conditions by tuning a single parameter, zone radius, it is not straightforward to choose the zone radius dynamically.

Hazy Sighted Link State (HSLS) protocol [53] is a link state protocol based on limited dissemination. Though HSLS does not *per se* have any reactive component as in ZRP, it partially exhibits behavior typical of reactive protocols, specifically use of suboptimal routes. Main idea here is to control the link state dissemination scope in space and time — closer nodes are sent link state updates more frequently compared to far away nodes. This idea is based on the observation that two nodes move slowly with respect to each other as the distance between them increases (also referred in the literature as the *distance effect*). So distant nodes through infrequent updates are only provided "hints" to route a packet closer toward the destination. As the packet approaches the destination, it takes advantage of progressively recent routing information that improve its chances of reaching the destination. A consequence of this limited dissemination strategy is that a packet may take suboptimal routes initially, but eventually arrives at the destination via an optimal route. Thus some amount of suboptimal routing is allowed to reduce the overall control overhead.

One important shortcoming of both ZRP and HSLS is that their design assumes a uniform traffic distribution and then optimizes the overall overhead. When the traffic is non-uniform, these protocols may not actually be efficient. A better strategy, perhaps, is to have the protocol also adapt to the traffic diversity.

3.6 Location-based Routing

Proactive, reactive or hybrid approaches we looked at in the previous sections share one common feature. In all these approaches, nodes discover (partial or full) topology information by exchanging routing messages and use this information to guide future routing decisions. We will now look at a completely different routing approach that utilizes *geographic location* of nodes.

Location-based (also called geographic) routing assumes that each node knows its own location by using the global positioning system (GPS) or some other indirect, localization technique. Besides, every node learns locations of its immediate neighbors by exchanging hello messages. The location of potential destination nodes is assumed to be available via a *location service*. When a source wants to send a packet to a destination, it uses the destination's location to find a neighbor that is closest in geographic distance to the destination, and closer than itself, and forwards the packet to that neighbor. That neighbor repeats the same procedure and until the packet makes it to the destination. This idea is often referred to as *greedy forwarding* in the literature. Note that greedy forwarding may fail to make progress, but we will postpone this discussion until later in this section.

Observe that geographic routing does not need any explicit route discovery or route maintenance mechanisms unlike other approaches. Except for gathering knowledge of node locations, nodes need not maintain any other routing state nor do they have to exchange any routing messages. As a result, geographic routing, in comparison with other approaches, can potentially be more efficient when topology changes quite frequently and can scale better with network size.

3.6.1 Location-based Routing Protocols

Location-Aided Routing (LAR) [30] is an optimization for reactive protocols to reduce flooding overhead. LAR uses an estimate of destination's location to restrict the flood to a small region (called *request zone*) relative to the whole network region. The idea is somewhat similar to query localization discussed in Section 4, although LAR was proposed earlier and it additionally demonstrates how location information can be exploited to benefit topology-based routing protocols.

LAR assumes that each node knows its own location, but does not employ any special location service to obtain location of other nodes. Destination location information obtained from a prior route discovery is used as an estimate of destination's location for limiting the flooding region in a subsequent route discovery. Two different LAR schemes with different heuristics to choose the request zone have been proposed. In the scheme that is shown to perform well, source floods a route request by including its estimate of destination's location and its estimated distance to destination in the request. Neighboring nodes receiving this request calculate their distance to destination using the destination's location in the request. If they are closer to the destination than the source, then they forward the request further by replacing the source's distance to destination with their own distance. A similar procedure is repeated at other nodes resulting in a directed flood toward the destination (Figure 3.3). Note that LAR uses location information only for finding routes and not for geographic forwarding of data packets.

Distance Routing Effect Algorithm for Mobility (DREAM) [1] is an early example of a routing protocol which is completely location-based. The location service is also part of the same protocol. With DREAM's location service, every node proactively updates every other node about its location. The overhead of such location updates is reduced in two ways. First, distance effect (nodes move

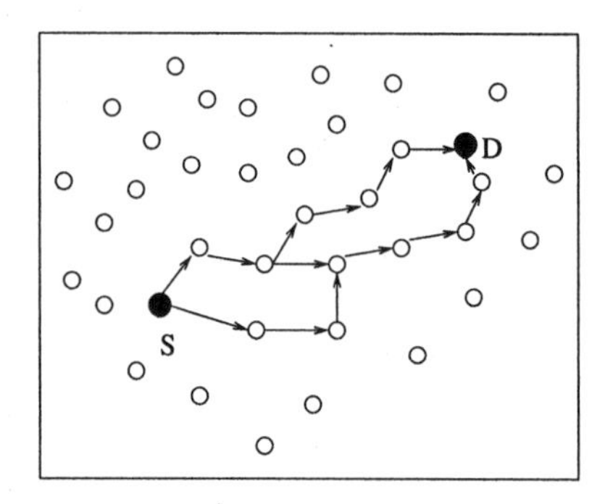

Figure 3.3. Restricted directional flooding in LAR and DREAM.

slowly with respect to each other as their distance of separation increases) is exploited by sending location updates to distant nodes less frequently than closer nodes (This is similar to HSLS (Section 5) which uses the distance effect for limited dissemination of link state updates). Second, each node generates updates about its location depending on its mobility rate — fast moving nodes update more often whereas slow moving nodes generate updates less often.

DREAM geographically forwards data packets in the form of a directional flood (similar to LAR's scheme for route request flood). Such directional flooding increases the likelihood of correct data delivery by compensating for inaccuracy in destination location information. At the same time, it can be very inefficient too. One simple mechanism to avoid this inefficiency is to follow the LAR strategy, except that here the first data packet acts as a route request; subsequent data packets will not be flooded, but they take the path found by the first packet.

Greedy Perimeter Stateless Routing (GPSR) [29] is a protocol that specifies only the geographic forwarding strategy, unlike DREAM, and assumes the existence of a location service. So any location service, either DREAM's location service or other schemes mentioned later in this section, could be used. GPSR's data forwarding algorithm comprises of two components: greedy forwarding and perimeter routing. Greedy forwarding is the same idea described at the beginning of this section. GPSR uses it as the default forwarding mechanism. But when greedy forwarding is not possible, perimeter routing is used. Greedy forwarding becomes impossible when the packet reaches a node which does not have any neighbor closer to the destination than itself, i.e., packet reaches a dead end or void. Figure 3.4 shows an example. The basic idea in perimeter routing is to begin at the node X where greedy forwarding failed and walk around the void until a node Y which is closer to the destination than X is reached. From then on (i.e., Y onward), greedy forwarding is resumed until the packet reaches the destination or another void is encountered. One might

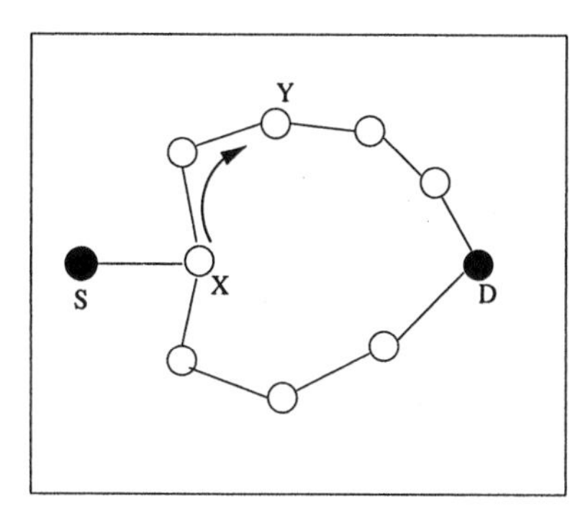

Figure 3.4. Illustration of greedy forwarding failure and perimeter routing in GPSR. In this figure, S is the source and D is the destination. By greedy forwarding, S sends the packet to node X. But all neighbors of X are farther to D than itself, so greedy forwarding fails at X. X then switches to perimeter mode and routes the packet along the perimeter until it reaches Y (closer to D than itself). From Y, greedy forwarding is used again until the packet reaches D. For simplicity, in this example we have assumed that actual network graph is planar.

wonder at this point why greedy forwarding is at all used if it can sometimes fail and why not simply use only perimeter routing always? The answer is as follows. Greedy forwarding is not only simple, but also optimal when it succeeds. On the other hand, perimeter routing always guarantees data delivery, it is seldom optimal.

In the simple description of perimeter routing above, we have omitted several key details. In order to apply perimeter routing, a planarized version of the actual graph has to be constructed first by removing all crossing edges. In simple terms, planar graph can be seen as collection of closed polygons stitched together. Local algorithms using only neighbors and their locations are available for constructing different kinds of planar graphs (e.g., restricted neighborhood graph and gabriel graph). Once a planar graph is constructed in a distributed manner, perimeter routing of a packet starting at a node X destined for node D reduces to moving across the successively closer faces to the destination which intersect the line segment joining X and D by traversing some edges in each face. Since perimeter routing moves across faces, it is also called *face routing*. Figure 3.4 illustrates the perimeter routing idea. Note that GPSR uses the planarized graph only for perimeter routing and the actual graph for greedy forwarding.

In a recent paper [33], it was observed that perimeter routing used in GPSR to come out of a void can result in much longer paths than needed. A new routing algorithm called *Greedy Other Adaptive Face Routing (GOAFR)* was also proposed in the same paper which avoids long paths by using a bounding region and a slightly different variant of perimeter routing. More importantly,

GOAFR is shown to be worst-case optimal and also on the average significantly outperforms GPSR. It is also worth mentioning here that most geographic routing protocols in the literature can be classified into greedy forwarding, face routing, or a combination of both.

3.6.2 Location Service Protocols

Location service providing destination's location is the key component of any system that does geographic routing. We already looked at one location service mechanism in the context of DREAM. Recall that in DREAM's mechanism every node maintains location information for every other node via a flooding-like (though optimized) location dissemination. So as the network size becomes large, the overhead of this mechanism grows very fast. Predictive mechanisms, such as *dead reckoning*, however, can be used to contain such overheads [34]. Here, infrequent dissemination may be sufficient if a movement model of the mobile nodes can be constructed.

At the opposite end, location service protocols can take a database approach. Location database systems typically rely on one or more nodes in the network that work as location servers. The servers may be dynamically elected. They are updated proactively by moving nodes. These systems are inefficient when locations are frequently queried, as this increases the query-reply load. *Grid Location Service (GLS)* [35] and *HomeZone* [55]) fall under this category.

In summary, geographic routing is undoubtedly a promising approach for large and dynamic networks, provided every node has the ability to find its own location and the availability of an efficient location service. Efficient location service in a mobile network is the key to the success of geographic routing because the location service amounts to a major fraction of the overhead. It is also important to recognize that geographic routing is still an evolving area with not sufficient evidence for substantial gains in overhead compared to traditional approaches. For instance, a recent performance comparison between DREAM and DSR has shown that DSR outperforms DREAM in both performance and efficiency in some scenarios [6]. Finally, a combination of geographic routing and local topology routing is worth investigating as a way to increase the likelihood of packet delivery in situations where the overhead of providing accurate location is high.

3.7 Concluding Remarks

In this chapter, we have described unicast routing protocols for mobile ad hoc networks – focusing on proactive, reactive and geographic approaches. In this process, we also reviewed efficient flooding techniques since most routing protocols employ some form of flooding. It is fair to conclude that no single routing approach is clearly superior to others in every possible scenario. Their relative merits are heavily dependent on the application context. Reactive protocols typically use a lower routing overhead when traffic diversity is low. But they incur a high route discovery latency and also may use suboptimal routes. So they may not be effective for delay-sensitive applications and short-lived flows. Hybrid protocols can potentially combine the benefits of both proactive and reactive approaches and avoid their drawbacks. However, more work is needed before this potential can be fully realized. Geographic routing has the potential to scale to large and dynamic networks. But to be effective, it needs some way to gather accurate location information for the current node, the destination node and also the neighboring nodes. The impact of inaccuracy in location information is not fully understood yet.

For the most part, we focused only on shortest-hop routing. Using load-aware metrics and protocols, instead will help better utilize the network resources by spreading the traffic uniformly across the network, especially in low mobility scenarios. Designing load-aware algorithms is, however, quite challenging because it not only requires an good understanding and modeling of wireless channel interference behavior, but also routing decisions and ensuing interference are intertwined.

A closely related issue is that of cross-layer interactions. Several performance studies have shown that cross-layer interactions can play a big part in determining overall network performance almost to the same extent as protocol mechanisms at each layer. This calls for a recognition of the sensitivity of a protocol's performance at one layer on the specific higher and lower layer protocols — choosing a different set of higher and lower protocols may give different performance results.

Finally in current research, routing protocols are designed in isolation by considering the wireless link as an uncontrollable entity, and abstracting out a "graph" view of the network and developing routing protocol for this graph. But the reality is different. Wireless link can indeed be tuned by varying certain transmission parameters. For example, increasing or decreasing transmission power can make or break a link. Likewise, varying the modulation and coding properties can change the quality of the link. Not exploiting this controllability of the wireless link limits the extent to which performance of a routing protocol can be optimized.

References

[1] S. Basagni, I. Chlamtac, V. R. Syrotiuk, and B. A. Woodward. A Distance Routing Effect Algorithm for Mobility (DREAM). In *Proceedings of IEEE/ACM MobiCom*, pages 76–84, 1998.

[2] D. Bertsekas and R. Gallager. *Data Networks*. Prentice Hall, 1992.

[3] D. M. Blough et al. On the Symmetric Range Assignment Problem in Wireless Ad Hoc Networks. In *Proceedings of IFIP Conference on Theoretical Computer Science*, pages 71–82, 2002.

[4] R. V. Boppana, M. K. Marina, and S. P. Konduru. An Analysis of Routing Techniques for Mobile and Ad Hoc Networks. In *Proceedings of International Conference on High Performance Computing (HiPC)*, pages 239–245, 1999.

[5] J. Broch, D. Maltz, D. Johnson, Y-C. Hu, and J. Jetcheva. A Performance Comparison of Multi-Hop Wireless Ad Hoc Network Routing Protocols. In *Proceedings of IEEE/ACM MobiCom*, pages 85–97, 1998.

[6] T. Camp et al. Performance Comparison of Two Location Based Routing Protocols for Ad Hoc Networks. In *Proceedings of IEEE Infocom*, pages 1678–1687, 2002.

[7] R. Castaneda, S. R. Das, and M. K. Marina. Query Localization Techniques for On-demand Protocols in Ad Hoc Networks. *ACM/Kluwer Wireless Networks*, 8(2/3):137–151, 2002.

[8] C. Cheng, R. Riley, S. P. R. Kumar, and J. J. Garcia-Luna-Aceves. A Loop-free Extended Bellman-Ford Routing Protocol Without Bouncing Effect. In *Proceedings of ACM SIGCOMM*, pages 224–236, 1989.

[9] T. Clausen et al. Optimized Link State Routing Protocol. http://www.ietf.org/internet-drafts/draft-ietf-manet-olsr-11.txt, July 2003. IETF Internet Draft (work in progress).

[10] S. R. Das, R. Castaneda, and J. Yan. Simulation-based Performance Evaluation of Routing Protocols for Mobile Ad hoc Networks. *ACM/Baltzer Mobile Networks and Applications (MONET)*, 5(3):179–189, 2000.

[11] R. Dube et al. Signal Stability-based Adaptive Routing Routing (SSA) for Ad Hoc Mobile Networks. *IEEE Personal Communications*, 4(1):36–45, 1997.

[12] E. Gafni and D. Bertsekas. Distributed Algorithms for Generating Loop-free Routes in Networks with Frequently Changing Topology. *IEEE Transactions on Communications*, 29(1):11–18, 1981.

[13] J. J. Garcia-Luna-Aceves. Loop-Free Routing Using Diffusing Computations. *IEEE/ACM Transactions on Networking*, 1(1):130–141, 1993.

[14] M. R. Garey and D. S. Johnson. *Computers and Intractability: A Guide to the Theory of NP-Completeness.* W. H. Freeman and Company, 1979.

[15] T. Goff, N. Abu-Ghazaleh, D. Phatak, and R. Kahvecioglu. Preemptive Routing in Ad Hoc Networks. In *Proceedings of IEEE/ACM MobiCom*, pages 43–52, 2001.

[16] S. Guha and S. Khuller. Approximation Algorithms for Connected Dominating Sets. *Algorithmica*, 20:374–387, 1998.

[17] Z. Haas, J. Halpern, and L. Li. Gossip-based Ad Hoc Routing. In *Proceedings of IEEE Infocom*, pages 1707–1716, 2002.

[18] Z. Haas and M. Pearlman. The Performance of Query Control Schemes for the Zone Routing Protocol. *IEEE/ACM Transactions on Networking*, 9(4):427–438, 2001.

[19] C. Hedrick. Routing Information Protocol. RFC 1058, 1988.

[20] G. Holland and N. Vaidya. Analysis of TCP Performance over Mobile Ad Hoc Networks. *ACM/Kluwer Wireless Networks*, 8(2/3):275–288, 2002.

[21] Y-C. Hu and D. Johnson. Caching strategies in On-demand Routing Protocols for Wireless Ad Hoc Networks. In *Proceedings of IEEE/ACM MobiCom*, pages 231–242, 2000.

[22] Y-C. Hu and D. Johnson. Implicit Source Routes for On-demand Ad Hoc Network Routing. In *Proceedings of ACM MOBIHOC*, pages 1–10, 2001.

[23] Y-C. Hu and D. Johnson. Ensuring Cache Freshness in On-demand Ad Hoc Routing Protocols. In *Proceedings of Int'l Workshop on Principles of Mobile Computing (POMC)*, pages 25–30, 2002.

[24] IEEE Standards Department. Wireless LAN Medium Access Control (MAC) and Physical Layer (PHY) Specifications, IEEE Standard 802.11–1997, 1997.

[25] J. M. Jaffe and F. H. Moss. A Responsive Distributed Routing Algorithm for Computer Networks. *IEEE Transactions on Communications*, 30(7):1758–1762, 1982.

[26] P. Johansson, T. Larsson, N. Hedman, B. Mielczarek, and M. Degermark. Scenario-based Performance Analysis of Routing Protocols for Mobile Ad-Hoc Networks. In *Proceedings of IEEE/ACM MobiCom*, pages 195–206, 1999.

[27] D. Johnson and D. Maltz. Dynamic Source Routing in Ad Hoc Wireless Networks. In T. Imielinski and H. Korth, editors, *Mobile computing*, chapter 5. Kluwer Academic, 1996.

[28] D. B. Johnson, D. A. Maltz, Y. Hu, and J. G. Jetcheva. The Dynamic Source Routing Protocol for Mobile Ad Hoc Networks (DSR).

http://www.ietf.org/internet-drafts/draft-ietf-manet-dsr-07.txt, Feb 2002. IETF Internet Draft (work in progress).

[29] B. Karp and H. T. Kung. GPSR: Greedy Perimeter Stateless Routing for Wireless Networks. In *Proceedings of IEEE/ACM MobiCom*, pages 243–254, 2000.

[30] Y. Ko and N. H. Vaidya. Location-Aided Routing (LAR) in Mobile Ad Hoc Networks. In *Proceedings of IEEE/ACM MobiCom*, pages 66–75, 1998.

[31] S. P. Konduru and R. V. Boppana. On Reducing Packet Latencies in Ad Hoc Networks. In *Proceedings of IEEE Wireless Communications and Networking Conference (WCNC)*, pages 1482–1487, 2000.

[32] U. C. Kozat, G. Kondylis, B. Ryu, and M. K. Marina. Virtual Dynamic Backbone for Mobile Ad Hoc Networks. In *Proceedings of IEEE International Conference on Communications (ICC)*, pages 250–255, 2001.

[33] F. Kuhn, R. Wattenhofer, and A. Zollinger. Worst-Case Optimal and Average-case Efficient Geometric Ad-hoc Routing. In *Proceedings of ACM MobiHoc*, pages 267–278, 2003.

[34] V. Kumar and S. R. Das. Performance of Dead Reckoning-Based Location Service for Mobile Ad Hoc Networks. *Wireless Communications and Mobile Computing*, 2003. To appear.

[35] J. Li, J. Jannoti, D. DeCouto, D. Karger, and R. Morris. A Scalable Location Service for Geographic Ad Hoc Routing. In *Proceedings of IEEE/ACM MobiCom*, pages 120–130, 2000.

[36] M. K. Marina and S. R. Das. On-demand Multipath Distance Vector Routing in Ad Hoc Networks. In *Proceedings of IEEE International Conference on Network Protocols (ICNP)*, pages 14–23, 2001.

[37] M. K. Marina and S. R. Das. Performance of Route Caching Strategies in Dynamic Source Routing. In *Proceedings of the Int'l Workshop on Wireless Networks and Mobile Computing (WNMC) in conjunction with Int'l Conf. on Distributed Computing Systems (ICDCS)*, pages 425–432, 2001.

[38] M. K. Marina and S. R. Das. Routing Performance in the Presence of Unidirectional Links in Multihop Wireless Networks. In *Proceedings of ACM MobiHoc*, pages 12–23, 2002.

[39] P. M. Merlin and A. Segall. A Failsafe Distributed Routing Protocol. *IEEE Transactions on Communications*, 27(9):1280–1287, 1979.

[40] J. Moy. OSPF version 2. RFC 2328, 1998.

[41] S. Murthy and J. J. Garcia-Luna-Aceves. An Efficient Routing Protocol for Wireless Networks. *ACM/Baltzer Mobile Networks and Applications*, 1(2):183–197, 1996.

[42] S.-Y. Ni, Y.-C. Tseng, Y.-S. Chen, and J.-P. Sheu. The Broadcast Storm Problem in a Mobile Ad Hoc Network. In *Proceedings of IEEE/ACM MobiCom*, pages 151–162, 1999.

[43] R. Ogier, M. Lewis, and F. Templin. Topology Dissemination Based on Reverse-Path Forwarding (TBRPF). http://www.ietf.org/internet-drafts/draft-ietf-manet-tbrpf-09.txt, June 2003. IETF Internet Draft (work in progress).

[44] V. D. Park and M. S. Corson. A Highly Adaptive Distributed Routing Algorithm for Mobile Wireless Networks. In *Proceedings of IEEE Infocom*, pages 1405–1413, 1997.

[45] M. Pearlman and Z. Haas. Determining the Optimal Configuration for the Zone Routing Protocol. *IEEE Journal on Selected Areas in Communications*, 17(8):1395–1414, 1999.

[46] W. Peng and X. Lu. On the Reduction of Broadcast Redundancy in Mobile Ad Hoc Networks. In *Proceedings of ACM MobiHoc*, pages 129–130, 2000. Poster.

[47] C. E. Perkins, E. Belding-Royer, and S. R. Das. Ad hoc On-Demand Distance Vector (AODV) Routing. http://www.ietf.org/rfc/rfc3561.txt, July 2003. RFC 3561.

[48] C. E. Perkins and P. Bhagwat. Highly Dynamic Destination-Sequenced Distance-Vector Routing (DSDV) for Mobile Computers. In *Proceedings of ACM SIGCOMM*, pages 234–244, 1994.

[49] C. E. Perkins and E. M. Royer. Ad Hoc On-Demand Distance Vector Routing. In *Proceedings of IEEE Workshop on Mobile Computing Systems and Applications (WMCSA)*, pages 90–100, 1999.

[50] C. E. Perkins, E. M. Royer, S. R. Das, and M. K. Marina. Performance Comparison of Two On-demand Routing Protocols for Ad Hoc Networks. *IEEE Personal Communications*, 8(1):16–28, 2001.

[51] A. Qayyum, L. Viennot, and A. Laouiti. Multipoint Relaying for Flooding Broadcast Messages in Mobile Wireless Networks. In *Proceedings of the 35th Annual Hawaii International Conference on System Sciences*, pages 3898–3907, 2002.

[52] C. Santivanez, B. McDonald, I. Stavrakakis, and R. Ramanathan. On the Scalability of Ad Hoc Routing Protocols. In *Proceedings of IEEE Infocom*, pages 1688–1697, 2002.

[53] C. Santivanez, R. Ramanathan, and I. Stavrakakis. Making Link-State Routing Scale for Ad Hoc Networks. In *Proceedings of ACM MobiHoc*, pages 22–32, 2001.

[54] R. Sivakumar, B. Das, and V. Bharghavan. Spine Routing in Ad Hoc Networks. *Cluster Computing*, 1:237–248, 1998.

[55] I. Stojmenovic. Home Agent Based Location Update and Destination Search Schemes in Ad Hoc Wireless Networks. Technical Report, Computer Science, University of Ottawa, September 1999.

[56] J. Sucec and I. Marsic. An Application of Parameter Estimation to Route Discovery By On-Demand Routing Protocols. In *Proceedings of IEEE ICDCS*, pages 207–216, 2001.

[57] C-K. Toh. Associativity-Based Routing for Ad-Hoc Mobile Networks. *Wireless Personal Communications*, 4(2):1–36, 1997.

[58] B. Williams and T. Camp. Comparision of Broadcasting Techniques for Mobile Ad Hoc Networks. In *Proceedings of ACM MobiHoc*, pages 194–205, 2002.

[59] J. Wu and F. Dai. A Generic Distributed Broadcast Scheme in Ad Hoc Wireless Networks. In *Proceedings of IEEE ICDCS*, pages 460–467, 2003.

Chapter 4

MULTICASTING IN AD HOC NETWORKS

Prasant Mohapatra, Jian Li, and Chao Gui
Department of Computer Science
University of California
Davis, CA 95616
prasant, lijian, guic@cs.ucdavis.edu

Abstract The widespread use of mobile and handheld devices is likely to popularize ad hoc networks, which do not require any wired infrastructure for intercommunication. The nodes of mobile ad hoc networks (MANETs) operate as end hosts as well as routers. They intercommunicate through single-hop and multi-hop paths in a peer-to-peer fashion. Most applications of MANETs require efficient support for multicast communications in which a node can communicate with multiple other nodes exploiting the broadcast nature of wireless channels. In this chapter, we first provide a classification approach of the mulitcasting techniques in mobile ad hoc networks, followed by the description of the protocols. Overarching issues such as energy efficiency, reliability, quality of service, and security have been also addressed. Several intriguing issues have been identified for further investigation on this topic.

Keywords: Mobile ad hoc networks, Multicasting, Broadcasting, Energy-efficient routing, Quality of service.

4.1 Introduction

The wireless mobile networks and devices are becoming increasingly popular as they provide users access to information and communication anytime and anywhere. The conventional wireless mobile communication is usually supported by a wired fixed infrastructure (like ATM or Internet). The mobile devices use single-hop wireless radio communication to access a base station that connects it to the wired infrastructure. In contrast, the class of mobile ad hoc networks (MANETs) does not use any fixed infrastructure. The nodes of MANETs intercommunicate through single-hop and multi-hop paths in a peer-to-peer fashion. Intermediate nodes between a pair of communicating nodes act as routers. Thus the nodes in MANETs operate both as hosts as well as routers. The nodes are mobile, and so the creation of routing paths is affected by the addition and deletion of nodes. The topology of the network may change rapidly and unexpectedly. Figure 4.1 shows an example of a mobile ad hoc network.

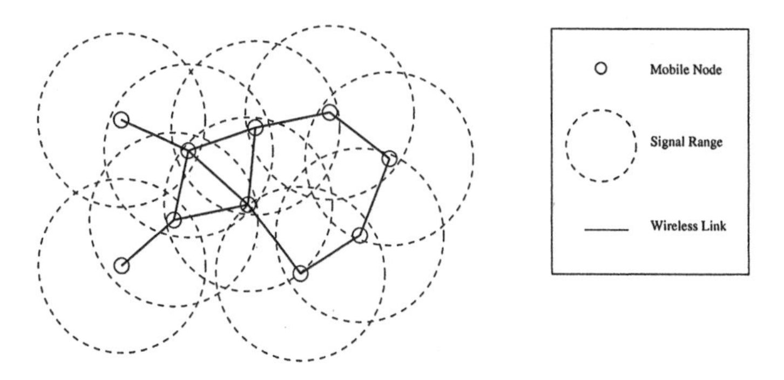

Figure 4.1. A mobile ad hoc network.

MANETs are useful in many application environments and do not need any infrastructure support. Collaborative computing and communications in smaller areas (buildings, organizations, conferences, etc.) can be set up using MANETs. Communications in battlefields and disaster recovery areas are other examples of application environments. Similarly communications using a network of sensors, and inter-island communications using floats over water are other potential applications of MANETs. The increasing use of collaborative applications and wireless devices may further add to the needs and usages of MANETs. Many of these potential applications of MANETs involve point-to-multipoint communication, and thus would benefit from multicasting support in the network layer.

Intercommunication in MANETs differs from that of wired networks in the following aspects.

- The wireless communication medium has variable and unpredictable characteristics. The signal strength and propagation delay fluctuate with respect to time and environment.

- The bandwidth availability and battery power are limited in mobile ad hoc networks. Thus the algorithms and protocols need to conserve bandwidth as well as energy.

- The computing components (processors, memory, I/O devices) used in wireless devices usually have low capacity and limited processing power. Thus the protocols for communications need to be lightweight in terms of computational and storage needs.

- The mobility of the nodes creates a continuously changing topology for communication. Routing paths break and new ones are formed dynamically.

- Unlike the wired network, wireless medium is a broadcast medium; all nodes in the transmission range of a node can hear the packets simultaneously.

In light of the above differences, the issues and challenges for intercommunication in MANETs are more complex than their wired counterpart.

IP multicasting was first proposed over a decade ago [1] as an extension to Internet architecture to support multiple clients at network layer. The fundamental motivation behind IP multicasting is to save network and bandwidth resource via transmitting a single copy of data to reach multiple receivers simultaneously. A basic principle for the forwarding tree is to branch as close to the receivers as possible. In ad hoc networks, we want to adhere to this requirement as closely as possible because of the severe bandwidth limitations in ad hoc networking environments.

Similar to Internet multicasting, it is necessary to deal with dynamic memberships in multicast groups in ad hoc networks. In both Internet and ad hoc multicasting, dynamic membership refers to the fact that individual clients may join and leave multicasting sessions dynamically. As a result, a multicast protocol needs to define operations of member join and leave, and how to recover from routing failure. The data forwarding path is constructed either as a tree or a mesh.

What makes ad hoc multicasting distinguished from Internet multicasting is that mobile nodes could move around freely and rapidly. In other words, we have to deal with high network dynamics due to node mobility, which makes ad hoc multicasting even more challenging. Ad hoc multicasting protocols in existing literature have either evolved from the Internet multicast protocol, or designed specifically for ad hoc networks. Most of these protocols attempt to adapt to the network dynamics in ad hoc networks. The primary goal of ad hoc multicasting protocols should be to construct/maintain a robust & efficient multicasting route even during high network dynamics. By "robust", we mean that the protocol should be able to operate correctly in spite of node mobility and topology changes. By "efficient", we mean both control and data forwarding overheads should be maintained low.

This chapter is organized as follows. In Section 4.2, we outline a classification methodology for multicasting protocols. The details of the specific protocols are described in Section 4.3. Broadcasting is discussed in Section 4.4. In Section 4.5, we provide a qualitative comparison of multicasting protocols. Overarching issues such as quality of service, energy conservation, reliability, and security are addressed in Section 4.6, followed by the concluding remarks in Section 4.7.

4.2 Classifications of Protocols

Multicasting techniques in MANETs can be classified based on group dynamics or network dynamics. In this section, we describe these two basis of classifications. Details of the protocols will follow in the next section.

4.2.1 Dealing with Group Dynamics

General principles for dealing with dynamic membership in ad hoc multicasting protocols are: on demand, receiver initiated, timer-based soft state. The basic idea of on demand approaches is to construct and maintain multicast routes only when needed. In receiver initiated approaches, it is the receiver's responsibility to find and keep track of a multicast session. If some states must be maintained to make a multicast session work, it is desirable to use timer-based soft state instead of hard state. Soft states are maintained on demand and refreshed from time to time; otherwise, its associated timer expires and the state is removed from intermediate nodes.

A primary issue for managing multicast group dynamics is the routing path that is built for data forwarding. Most existing ad hoc multicasting protocols can be classified as tree-based or mesh-based. In a tree-based protocol, a tree-like data forwarding path is built with the root at the source of the multicast session. The multicast tree consists of a unique path from the sender to a receiver. This can be extended to a shared-tree when multiple multicast sessions are in parallel in the network and can share some common parts of data forwarding trees with each other. In a mesh-based protocol, in contrast, multiple routes may exist between any pair of source and destination, which is intended to enrich the connectivity among group members for better resilience against topology changes.

A major difference between tree-based and mesh-based protocols lies in the manner in which a multicast message is relayed. In a tree-based protocol, each intermediate node on the tree has a well-defined list of next hops for a specific multicast session. It will send a copy of the received message to only the neighboring nodes on its nexthop list. In most mesh-based protocols, however, relaying transmission takes a more redundant approach: each node on the mesh will broadcast the message upon its first reception of the message. Although this transmission redundancy may lead to higher overheads in many cases, it is still worth because of its resilience against dynamic topology and link quality.

A compromise between the tree-based and mesh-based protocols is made in MCEDAR [2] builds a mesh structure among multicast members to obtain redundant connectivity, and extract a data forwarding tree on top of the mesh structure.

To the best of our knowledge, all the existing protocols for ad hoc multicasting, both tree-based and mesh-based, do not make explicit efforts to take advantage of the broadcast nature of wireless medium. In an ad hoc network with a shared wireless channel, a link between any pair of nodes is not well defined as in the case of Internet. Instead, any two nodes within each other's transmission ranges may form a link, and a node may form multiple links to its neighbors simultaneously. We believe this feature of broadcast media can help in reducing transmission overhead if we take it into consideration when constructing the multicast routes.

4.2.2 Dealing with Network Dynamics

As mentioned earlier, we need to overcome the network dynamics in order to achieve robust and efficient ad hoc multicasting. A major source of network dynamics is node mobility and node failure. In this subsection we summarize some basic approaches to addressing this challenge in existing literature. We cite some examples for each of the approaches.

Approach 1: Reliance on More Nodes Since nodes are mobile, every intermediate node could be a possible cause of route breakage. If we include more nodes in the multicast infrastructure, we can obtain better connectivity among group members. When a link breaks due to node mobility, we may have a good chance to find an alternative route. In other words, we don't need to initiate route maintenance procedure frequently corresponding to every single link failure.

The second advantage of this approach is related to the transmission aspect: redundant transmissions can offset the influence of the unreliable wireless links. In many mesh-based protocols, within-mesh flooding is a common choice to improve reliability of data delivery.

Some examples for this approach include CAMP's forwarding group [3], ODMRP [4], and neighbor support multicasting [5].

Approach 2: Reliance on Fewer Nodes Since nodes are mobile, it is time- and resource-consuming for a large number of nodes to get involved in route construction and maintenance. By limiting the number of nodes involved, the control overhead can be reduced. We can extract a virtual backbone, typically a dominating set of the entire network, and rely on the backbone while constructing multicast route when a new session starts.

MCEDAR [2] uses this approach for ad hoc multicast. This approach is also used in unicast ad hoc routing [6] [7].

Approach 3: Reliance on No Nodes Since all nodes are mobile, the multicast routes would need maintenance as time goes by. If session states are stored in packet headers, the protocol does not have to rely on any specific nodes to form a forwarding path because we do not even need one! Session states carried in packet header may be a list of node IDs, or a series of location coordinates. In a protocol using this approach, intermediate nodes check the packet header and decide where or who to forward the packet.

Examples for this approaches include location guided small group communication [8] and DDM [9].

Approach 4: Reliance on Stabler Nodes This approach attempts to take advantage of node mobility and network architecture. If nodes in a network have different degrees of mobility, for example, fast and slow, we can rely on the slow (thus stabler) nodes in order to build the multicast route. Note that "fast" and "slow" could be in the relative sense. For example, a group of nodes may move fast together toward a common direction, but the relative speed among group member is "slow".

An example of this approach is M-LANMAR [10], which is a multicast protocol exploiting team-based mobility.

Approach 5: Reliance on an Overlay Layer Since all nodes are mobile, adapting to network dynamics is an extra burden for multicasting protocols. By inserting a middle layer in between, we can hide dynamics in lower layer and let multicast protocols concentrate only on multicasting. In this approach, the protocols normally build an overlay mesh on top of the physical network, and the multicasting route is built on top of this overlay mesh. Without knowing the underlying dynamics, it is easier for the multicast protocol to focus on implementing multicast functionalities.

Protocols using this approach includes AMRoute [11] and PAST-DM [12], both of which construct a virtual mesh structure on top of the physical network. The virtual mesh relies on some unicast routing protocols to provide tunneling route between any two nodes on mesh. Data forwarding tree is extracted on top of the virtual mesh, and is unaware of underlying topology changes.

Several popular multicasting protocols are classified according to our discussion and are summarized in Table 4.1.

4.3 Multicasting Protocols

In the previous sections, we reviewed the special properties of mobile ad hoc networks, and examined how these properties affect the design and implementation of network protocols. To deliver packets effectively to the multicast group members, any multicasting protocol should address these properties. In

Protocol name	Data forwarding infrastructure	Approach to dealing with network dynamics
AMRoute	tree on top of mesh	rely on a middle layer
CAMP	mesh	rely on more nodes
ODMRP	mesh	rely on more nodes
MCEDAR	tree on top of mesh	rely on fewer nodes
MAODV	tree	-
M-LANMAR	tree	rely on stabler nodes
DDM	tree	rely on no nodes
Location Guided Multicast	tree	rely on no nodes

Table 4.1. Classification

this section, we present several multicasting protocols proposed specifically for the mobile ad hoc networks.

4.3.1 Multicast operations of AODV (MAODV)

As the multicast protocol associated with AODV [13], MAODV [14] uses the conventional tree-based approach for multicast routing. Besides the routing table, each node maintains a Multicast Route Table (MRT) to support multicast routing. A node adds new entries into the MRT after it is included in the route for a multicast group. Each entry records the multicast group IP address, group leader IP address, group sequence number and next_hops (neighbors on the multicast tree).

Each multicast group also needs its own sequence number in order to indicate the freshness of a multicast route, which is maintained by the group leader. When a node wishes to join a multicast group and it does not know who is the leader, it broadcasts a RREQ packet with destination field set as the group ID address. If it does not receive a RREP before timing out, it will retry for certain number of times. Subsequent unsuccessful attempts would mean that there are no other members of the group within its connected portion of the network. In such cases, it assumes the group leadership. It initializes the group sequence number to one, and broadcasts a Group Hello packet across the network periodically with step-wise incremented sequence number.

Every node keeps record of who is the leader of which group by promiscuously listening to RREPs. Thus, if it want to join a group, it may have the address of the leader. If it also has a route to the leader in its routing table, it can unicast the join RREQ to the leader directly. Otherwise, it will broadcast the join RREQ packet. If a member node loses its route to the group, it broadcasts a normal RREQ when it want to send data to the group.

If a node receives a join RREQ, it can reply if it is a router on the group's multicast tree and it holds a group sequence number that is high enough, while the group leader always can reply join RREQ. RREP is unicasted, and the responding node updates its MRT accordingly. RREP contains the last known group sequence number, address of group leader, and a special field called Mgroup_Hop. This field is initialized to zero. When a node on the path to the source node receives the RREP, it increases its Mgroup_Hop field, and updates to its multicast route table. When the source node receives the RREP, it can determine the hop distance to the nearest router on the group's tree, and a new branch of the tree is also built at the same time. Moreover, the whole multicast tree is gradually built up while branches are added one by one. When a node on the tree receives a packet targeting its group address, it will multicast the packet to all its neighbors on the tree. To ensure loop-free property, it is necessary to make sure only one router on the tree responds the join RREQ. If multiple responses do arrive, the source node should accept only one. All the other responses will be ignored and finally invalidated by expiration timers.

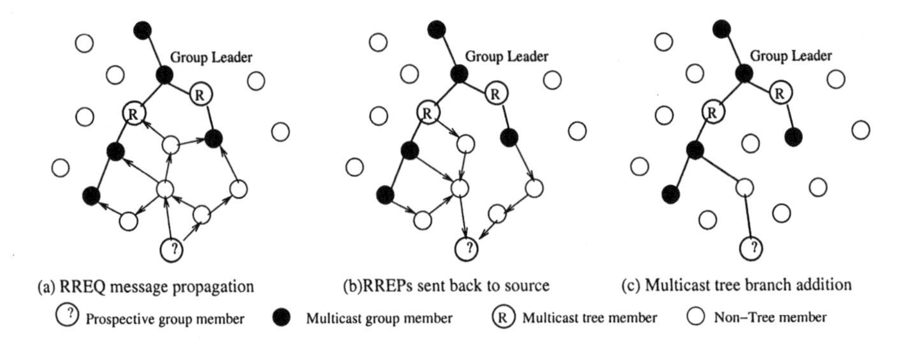

Figure 4.2. Multicast join operation of MAODV.

When a member decide to leave its group, and if it is not a leaf node in the multicast tree, it must continue to serve as a router. If it is a leaf node, it will have only one immediate neighbor. The node unicast a leave message to that neighbor and clears all information about the group in its tables. The neighbor, upon receiving the message will update its neighbor list as well. If it is not a member and it becomes a leaf node after the pruning, it will start its own pruning by doing the same. So the pruning stops when either a group member or a non-leaf node is reached.

Multicast tree links may break due to node mobility or timer expiration, and this will be detected by both end nodes of the link. But only the downstream node will be responsible for the repair. To repair, it broadcasts a join RREQ with destination address set as group leader and Mgroup_Hop field set to its distance from the leader. The last known group sequence number is also included. To restrict the effects, the TTL field of the RREQ is set to a small value. If no reply is received before the time out period, the retrials will be network wide broadcasts. The nodes that can respond to this RREQ are those that are at least as close to the group leader as indicated by the packet. This prevents those nodes on the same side of the broken link from responding. Finally, when RREPs are unicasted to the initiating node, the procedure is the same as the joining of a new node.

If no RREP is received, it is assumed that the network has been partitioned and the tree cannot be reconnected. The partition of the tree that is downstream of the broken link can select a new leader. It must be a group member, and the new leader will distribute new round of group sequence numbers to its members. Later on, the partitions of network may become connected. Then, the two leaders will know each other since they are both broadcasting group hello messages, and will negotiate and combine there partitions and one leader will stop its role.

4.3.2 Reliance on More Nodes

Node mobility poses a great challenge to multicast routing. It is mandatory that a tree-based routing protocol should continuously update itself to accommodate the changing network topology. When a tree link breaks, a branch of the multicast tree becomes disconnected. The routing protocol needs to reconnect the partitioned branches swiftly. In a highly mobile network, the robustness and efficiency would be an important issue. If we allow path redundancy in the routing structure, i.e., allow the presence of multiple paths between certain node pairs, we change the tree structure into a mesh structure. Thus, the routing structure does not need to react to every link breakage, which is a nice feature especially in the mobile ad hoc network settings. However, the path redundancy will reduce the data forwarding efficiency, and there could be loop formations. Thus, a mesh-based routing protocol needs a very careful design.

4.3.2.1 Core-Assisted Mesh Protocol (CAMP). CAMP [3] is designed to support multicast routing in very dynamic ad hoc networks using a shared mesh structure. It ensures that the shortest paths from all receivers to the sources (called reverse shortest paths) are included in the group's mesh. Figure 4.3 illustrates how data packets are forwarded from router h to the rest of the group members in CAMP and in a shared-tree multicast protocol. To prevent packet replication or looping in the mesh, each node maintains a cache to keep track of recently forwarded packets. Periodically, a receiver node reviews its packet cache in order to determine whether it is receiving data packets from those neighbors which are not on the reverse shortest path to the source. Whenever such situation arises, a *heartbeat* message is sent to successor in its reverse shortest path to the source. That *heartbeat* message triggers a *push join* message when the successor is not a mesh member. This procedure ensures all the nodes along any reverse shortest path are included in the mesh.

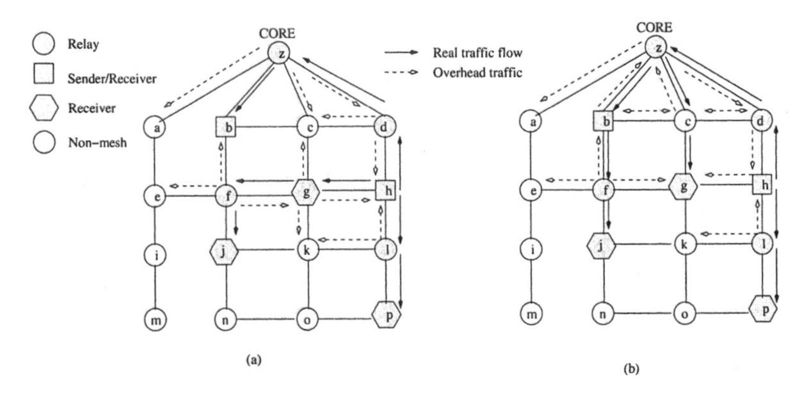

Figure 4.3. Traffic flow from h. (a) In a CAMP mesh. (b) In the equivalent shared tree.

CAMP uses core node for limiting the control traffic needed for creation of multicast meshes. Unlike CBT, it does not require that all traffic should flow through the core nodes. If a node wishing to join a multicast group finds that it has neighbors which are duplex members of the group, it simply updates its multicast routing table (MRT) and announces its membership to the neighbors using a standard multicast routing update procedure. When none of its neighbors are mesh members, it either sends a join request toward a core or attempts to reach a group member by the expanding ring search. Any duplex member of the mesh can respond a join request with a join ACK, which is propagated back to the originator of the request. The normal mesh members are called duplex nodes. Besides, CAMP allows a node to join the mesh in a simplex mode when creating one-way connections between sender-only nodes and the rest of the multicast mesh.

4.3.2.2 On-Demand Multicast Routing Protocol (ODMRP).

ODMRP [4] [15] extends the concept of mesh with the *forwarding group* concept. The forwarding group is a set of nodes responsible for forwarding multicast data on shortest paths between any member pairs, as is shown in Figure 4.4.

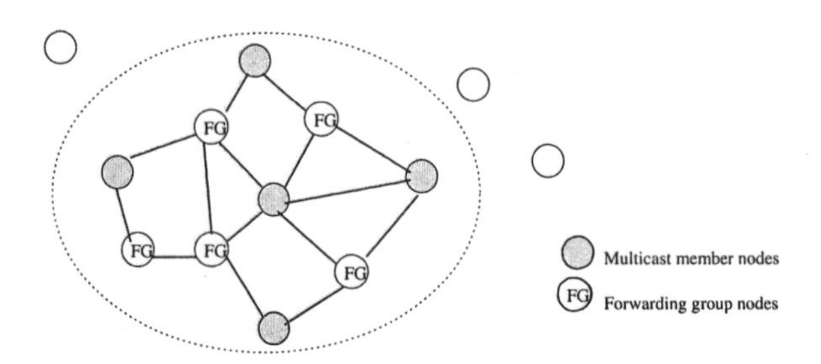

Figure 4.4. The forwarding group concept.

In ODMRP, group membership and multicast mesh are established and updated by each source *on demand.* By flooding a JOIN Query, a source node starts building a forwarding mesh for the multicast group, and collect membership information at the same time. When a node receives a non-duplicate JOIN Query, it stores the upstream node ID and rebroadcasts the packet. When the JOIN REQUEST packet reaches a multicast receiver, the receiver creates or updates the source entry in its *Member Table.* A JOIN Reply packet is then prepared and broadcasted by the receiver node. The packet is relayed back towards the source along the reverse path traversed by the JOIN Query packet. This process constructs (or updates) the routes from sources to receivers and builds a mesh of nodes, the *forwarding group.* Multicast sources refresh the membership information and update the routes by sending JOIN Query periodically.

8 bits	8 bits	8 bits	8 bits
Type	Reserved	Time To Live	Hop Count
Multicast Group Address			
Sequence Number			
Source IP Address			
Previous Hop IP Address			

Figure 4.5. Format of JOIN Query packet.

Figure 4.5 shows the format of a JOIN Query packet. When a multicast source has data packets to send but no route is known, it originates a "Join Query" packet. The source set Type field to 01 (which means a Join Query packet). Hop Count is initially set to zero. The TIME_TO_LIVE value for the packet should be adjusted based on network size and network diameter. The Sequence Number must be large enough to prevent wraparound ambiguity. When a node receives a Join Query packet, the following process is adopted.

1 Check if it is a duplicate by comparing the (Source IP Address, Sequence Number) combination with the entries in the message cache. If a duplicate, then discard the packet. DONE.

2 If it is not a duplicate, insert an entry into the message cache with the information of the received packet (i.e., sequence number and source IP address) and insert/update the entry for routing table (i.e., backward learning).

3 If the node is a member of the multicast group, it originates a Join Reply packet with the RET value enclosed.

4 Increase the Hop Count field by 1 and decrease the TTL field by 1.

5 If the TTL field value is less than or equal to 0, then discard the packet. DONE.

6 If the TTL field value is greater than 0, then set the node's IP Address into Last Hop IP Address field and broadcast. DONE.

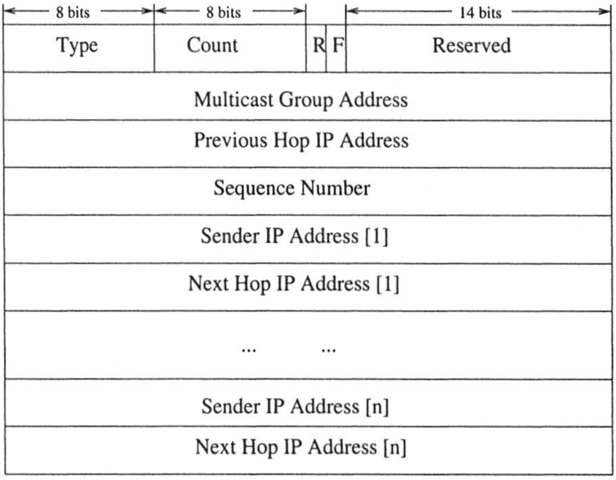

8 bits	8 bits			14 bits
Type	Count	R	F	Reserved
Multicast Group Address				
Previous Hop IP Address				
Sequence Number				
Sender IP Address [1]				
Next Hop IP Address [1]				
... ...				
Sender IP Address [n]				
Next Hop IP Address [n]				

Figure 4.6. Format of JOIN Reply packet.

A multicast receiver transmits a "Join Reply" packet after selecting the multicast route. Each sender IP address and next hop IP address of a multicast group are contained in the Join Reply packet. Figure 4.6 shows the format of a JOIN Reply packet.

When a Join Reply is received:

1 The node looks up the Next Hop IP Address field of the received Join Reply entries. If no entries match the node's IP Address, do nothing. DONE.

2 If one or more entries coincide with the node's IP Address, set the FG_FLAG and build its own Join Reply. The next hop IP address can be obtained from the routing table.

3 Broadcast the Join Reply packet to the neighbor nodes. DONE.

One salient feature of ODMRP is the *soft state* approach in maintaining multicast group members. For each member, the group membership is periodically renewed by the rounds of request/reply procedure. Once a member wants to leave a group, no additional signaling is needed. It simply just stops responding to the Join Query packets. This feature is very suitable for the mobile ad hoc network environment, in which join/leave operations may happen more frequently, and the cost of group maintenance is very high.

4.3.3 Reliance on Backbone Structure

Backbone-based multicast uses a hierarchical routing technique. The multicast routing is divided into two levels: the global multicast routing within the virtual backbone, and the local multicast from each backbone node to its dominated receiver nodes.

For a backbone-based approach, a distributed election process is conducted among all nodes in the network, so that a subset of nodes are selected as core nodes. The topology induced by the core nodes and paths connecting them form the virtual backbone, which can be shared by both unicast and multicast routing. In MCEDAR [2], a distributed *minimum dominating set (MDS)* algorithm is applied for this purpose, and the resulting backbone has the property that all nodes are within one hop away from a core node. A core node and its dominated node set form a cluster. The protocol proposed in [16] uses a more complex selection process. It relaxes the dominating set property by allowing each backbone node to be a root of a *local group* involving all nearby nodes. It is assumed in [17] that the "slow node" population is dense enough to form a connected spanning topology of the whole network area. Thus, the backbone is built upon the "slow nodes".

Once a virtual backbone is formed, the multicast operation is divided into two levels. The lower level multicast, which is within a cluster, is trivial. For the upper level multicast, the protocol in [16] uses a pure flooding approach within the backbone. MCEDAR builds a routing mesh, named as *mgraph*, within the virtual backbone, to connect all core nodes.

4.3.3.1 Multicast Core-Extraction Distributed Ad hoc Routing (MCEDAR). The protocol bases its *core-extraction* criteria on the *dominating set* concept for the network topology, namely, a node is either a core or an immediate neighbor of a core. Besides the adopted virtual backbone procedures, the protocol itself has four key components: (1)the mesh-based multicast structure, (2) the join protocol, (3) the core broadcast based forwarding protocol, and (4) the leave pruning and reconstruction protocols.

MCEDAR uses a mesh structure called the *mgraph* as the multicast routing structure. Only the core nodes can become members of an *mgraph*. Thus, the number of nodes involved in a multicast routing structure is significantly reduced. Each member of an *mgraph* maintains a notion of which of its nearby core nodes are members of the same *mgraph*. This information is used in data forwarding. Each *mgraph* is also associated with a robustness factor R.

When a node wants to join a group, it requires its dominating core to join the appropriate *mgraph*, and the dominator then performs the join operation. In order to prevent loop formations during *mgraph* reconstructions, each member also maintains a notion of local ordering among each other. Thus, each *mgraph* member is assigned a value named as JoinID, which has an initial value as infinity, and is updated during the course of *mgraph* construction.

The new joining core broadcasts a *join request JOIN(joinID)*. The *joinID* for a fresh joining core is set as infinity. The *join request* is relayed further by the non-member nodes in accordance to the core broadcast procedure. When a group member receives a join request, it sends back a *JOIN-ACK(joinID)* if its joinID is less than the joinID field of the arriving join request. When an intermediate core on the reverse path receives a JOIN-ACK, it relays the packet only when its number of neighbors in the *mgraph* does not exceed the robustness factor, R. Before relaying the JOIN-ACK, the intermediate core updates its own joinID if the relayed JOIN-ACK has lower joinID value, otherwise, it updates the joinID field of the JOIN-ACK packet. Figure 4.7 shows an example of the join procedure. The left-hand side shows how the *join request* is relayed up to the nodes in the core graph. The right-hand side shows the reverse paths for JOIN-ACKs and the update of joinIDs at each relaying nodes.

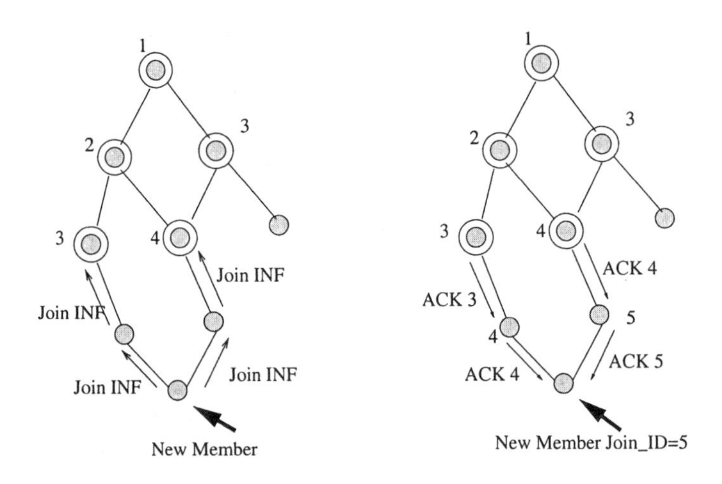

Figure 4.7. MCEDAR join procedure.

The forwarding protocol of MCEDAR follows the core broadcast procedure. When a data packet arrives at an *mgraph* member, the member attempts to forward the packet only to those nearby cores on the same *mgraph*. MCEDAR introduces a core broadcast procedure, which is used for a core node to flood a message to all the other core nodes in the network. The procedure is more efficient than the normal hop-by-hop flooding. The core broadcast procedure implicitly creates a source based tree that represents the fastest delivery structure for each source of the group.

A member of the *mgraph* issues a leave message only when the following two conditions hold. (1) It does not have any local members in its domain, and (2) its child list is empty. A leaving member needs to issue the message to all its parents.[1] A parent that receives the leave message from one of its children deletes the corresponding child's ID from its child list.

In some cases, an *mgraph* member can issue a reconstruction request to the other members. In order to prevent loop formations during this process, the JoinID field at each node is used in the following manner. A member issues a reconstruction request only when it looses connectivity with all its neighboring *mgraph* members who has smaller join times, i.e., JoinID values. A member responds to a reconstruction request only if its join time is less than the join time of the originator of the request.

4.3.4 Stateless Multicasting

In the *stateless multicast* protocols, the forwarding states are included in packet header, and no protocol state is maintained at any nodes except for the source node. From the information included in the packet headers, any intermediate node knows how to forward or duplicate the packet. Although packing routing information together with data traffic will enlarge data packet size, it reduces the total number of control packets generated by the protocol. Besides, when the group is idle, there is no control overhead.

4.3.4.1 Differential Destination Multicast (DDM). DDM [9] is intended for small group multicast. It not only adopts the stateless approach, but may also operate in a *soft state* mode. In this mode, intermediate nodes cache the forwarding states read from the packet header. The protocol no longer needs to list all destinations in every data packet header. When changes occur, an upstream node only needs to inform its downstream neighbors regarding the difference in destination forwarding since the last packet.

In DDM, the multicast data packets contain a payload and a DDM header, which is composed of a list of DDM blocks. Each DDM block is constructed for a particular downstream neighbor. Each DDM block contains the *intended receiver*, the *DDM block type*, the *block sequence number* and some other fields depending on the type. There are three types of the DDM blocks: Empty (E) blocks, Refresh (R) blocks and Difference (D) blocks. Except for the E block, both R block and D block have a destination list L. When used in broadcast media networks, DDM blocks for different downstream neighbors may be aggregated together into the header of one data packet. When the intended neighbors

[1] Since a *mgraph* is a mesh structure, each member can have multiple parents. The hierarchy is derived from the JoinID field of each node.

receive the data packet, each neighbor can locate the correct DDM block and the destination list for itself.

Each node maintains one Forwarding Set (FS) for each active multicast session. It records to which destinations this node needs to forward multicast data. When a node receives a DDM data packet from an upstream neighbor, it first locates the DDM block intended for itself, and check its block sequence number to see if it a duplicate one just seen before. The receiver then updates its FS according to the DDM block type. For an R block, the subset of destinations in FS which are cached from previous received DDM block from the same upstream node are totally replaced by the list in the newest DDM block. For a D block, that subset is incremented or decremented by the list in the newest DDM block. For an E block, that subset is removed from the FS. When the FS is updated, the destinations in the FS may be reached via different paths. Therefore, the FS is further partitioned into subsets according to the next hops. Thus, a new DDM header, containing new DDM blocks, is prepared for the next transmission of the data packet.

DDM follows the 1-to-n communication model. The source acts as an admission controller for the information it sends. New members join the group by unicasting a JOIN REQUEST to the source. DDM relies on the unicast protocol to quickly provide the next hop for any destination, which may be a hard request for the on-demand type unicast protocols.

4.3.5 Overlay Multicasting

Overlay multicast [18] [19] [20] has been proposed as an alternative approach for providing multicast services in the Internet. A virtual infrastructure can be built to form an overlay network on top of the physical Internet. Each link in the virtual infrastructure is a unicast tunnel in the physical network. The IP layer provides a best-effort unicast datagram service, while the overlay network implements all the multicast functionalities such as dynamic membership maintenance, packet duplication and multicast routing. AMRoute [11] is an ad hoc multicasting protocol that uses the overlay multicast approach. Bidirectional unicast tunnels are used to connect the multicast group members into a virtual mesh. After the mesh creation phase, a shared tree is created for data delivery, and is maintained within the mesh. One member node is designated as the logical core, which is responsible for initiating the tree creation process periodically. Figure 4.8 illustrates the concept of virtual mesh and the shared tree built by the AMRoute protocol within the mesh.

When the overlay multicasting technique is applied to the MANETs, the manner in which the overlay layer interacts with the physical network is quite different from that of overlay multicasting in the Internet. In MANETs, each node acts as a router as well as an end host. In most cases, we can assume the

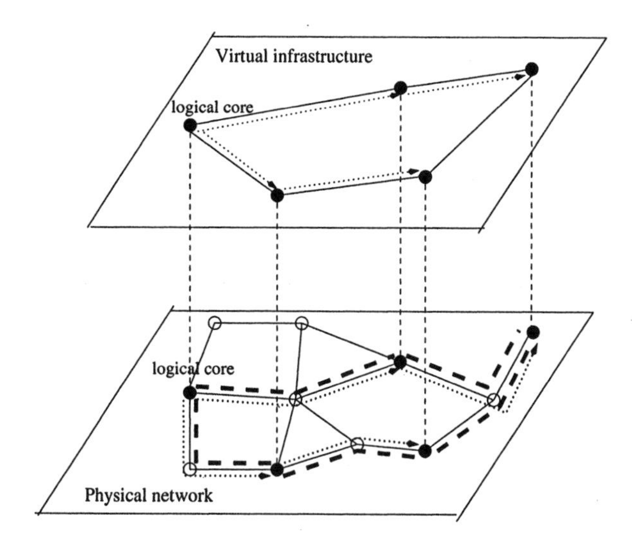

Figure 4.8. Concept of virtual topology for overlay multicast.

bandwidth homogeneity among the nodes in a MANET topology. Whereas in the Internet topology, there is a significant difference in available bandwidth at the end hosts and the routers. Forwarding and duplicating packets at the bandwidth limited endhosts are inherently less efficient than at the routers. Thus, there is a major efficiency problem in overlay multicasting in the Internet, compared to the network layer multicast. However, this problem does not exist for overlay multicasting in MANET.

4.3.5.1 Ad-hoc Multicast Routing Protocol(AMRoute). AMRoute [11] creates a per group multicast distribution tree using unicast tunnels connecting group members. Each group in the network has at least one logical core that is responsible for discovering new group members and creating/maintaining the multicast tree for data distribution. There are two main phases in the protocol operations: mesh creation and tree creation. Figure 4.8 illustrates the AMRoute overlay topology and the logical core.

It is much simpler to maintain a mesh than a tree at the member mutual discovery phase. Initially, each group member declares itself as a core for its own group of size one. Each core periodically floods JOIN-REQ messages with increasing TTL to discover other disjoint mesh segments for the group. When a member node receives a JOIN-REQ from a core of a different mesh segment for the same group, the node responds back with a JOIN-ACK. A new bi-directional tunnel is established between the core and the responding node of the other mesh segment. Due to mesh segment mergers, a mesh segment will have more than one cores. One of the cores will emerge as the "winning" core of the unified mesh due to the core resolution algorithm.

The core is responsible for initiating the tree creation process, which identifies the subset of the links within the mesh to form the shared data delivery tree. The core sends out periodic TREE-CREATE messages along all the links incident on it in the mesh. Group members receiving non-duplicate TREE-CREATE messages forward them on all mesh links except the incoming, and mark the incoming and outgoing links as tree links. If a duplicate TREE-CREATE message is received, a TREE-CREATE-NAK is sent back along the incoming links, which makes both end nodes of the mesh link mark it as mesh link instead of a tree link.

AMRoute operates independent of the underlying unicast protocol. This independence allows use of the optimal ad hoc unicast protocol for the network and can work transparently across domains supporting different unicast protocols.

4.3.6 Location Aided Multicasting

In networks where Global Positioning System (GPS) is available, each node is provided with the location and mobility information. The location aided routing techniques are utilized by the unicast protocols. The multicast protocols can also utilize this information for improving protocol robustness, or even making forwarding computation.

With GPS support, ODMRP [4] can be made adaptive to node movements by utilizing mobility prediction. By using location and mobility information, route expiration time can be estimated and receivers can select the path that will remain valid for the longest duration. With the mobility prediction method, sources can reconstruct routes in anticipation of route breaks. Thus, the protocol can be more resilient to node mobility.

Location Guided Tree Construction Algorithm for Small Group Multicast (LGT) [8] is also an overlay multicast protocol, where multicast data is encapsulated in a unicast packet and transmitted among group members. Using the location information of the group member nodes, the multicast tree is constructed without the knowledge of the network topology. The authors propose two types of heuristics, namely the location-guided k-array tree (LGK), and the location-guided Steiner tree (LGS) to construct the multicast tree with location information.

4.3.7 Gossip-Based Multicasting

Gossip, as a form of probabilistically controlled flooding, has been used to solve a number of problems such as network news dissemination. The basic idea of applying gossip to multicasting is to have each member node periodically "talk" to a random subset of other members. After each round of talk, the both gossipers can recover their missed multicast packets from each other. In contrast to deterministic approaches, probabilistic schemes will better survive a highly dynamic ad hoc network since it is independent of network topology, and its nondeterministic property matches the network characteristics.

Anonymous Gossip (AG): AG [21] is a multicast performance enhancement technique applied on top of any of the tree-based and mesh-based protocols with very little overhead. It is called anonymous gossip because it does not require a group member to have any knowledge of the other group members. A multicast protocol based on anonymous gossip would proceed in two phases. In the first phase, packets are multicast to the group using any unreliable multicast protocol. In the second phase, periodic anonymous gossip takes place in the background for each group member to recover any lost data packet from other members of the group that might have received it.

Route Driven Gossip (RDG): RDG [22] relies on a unicast protocol such as DSR [23] to provide routing information, which is used for guiding the gossip process. Each node maintains the following data structures for a multicast group: a data buffer which stores data packets received, and a *view* which is a list all other group member nodes known to this node. The view at each node is divided into two parts: *active view*, which contains the IDs of known members to which at least one routing path is known, and *passive view* which contains the IDs of known member to which no routing path is currently available. A node intending to join a group floods the network with a Group-Request message. All members receiving the message update their *active view*. They also return a Group-Reply to the request initiator with a certain probability. The initiator also updates its *active view* after receiving the Group-Reply.

Each member node periodically generates a gossip message and gossips it to a set of other nodes randomly chosen from its *active view*. The message includes a selected subset of the data buffer, and the sequence number of the most recent missing data packets. A group member receiving a gossip message will update its view of other group members and update its data buffer with newly received data. In responding the gossip initiator's request of recovering the missing data, the receiving node will unicast the missing data back to the initiator if the data is in its data buffer.

4.4 Broadcasting

Network wide broadcasting is an important function in MANETs, which attempts to deliver packets from a source node to all other nodes in the network. Broadcasting is often used as a building block for route discovery in on-demand ad hoc routing protocols.

For designing broadcast protocols for ad hoc networks, one of the primary goal is to reduce the overhead (collision and retransmission, redundant retransmission, etc.) while reaching all the nodes in the network. The implication of the broadcast nature of wireless signal is twofold. In one view, it would cause more contentions and collisions in the shared wireless channel. However, from another viewpoint, it provides the capability of reaching multiple neighboring nodes via a single transmission. In wireless broadcasting, if all neighboring nodes relay (rebroadcast) the received packet immediately, it will result in the problem of "broadcast storm". To avoid the broadcast storm problem, some form of randomized delay can be introduced before a neighboring node relays the received packet.

With the support from MAC layer using RTS/CTS/DATA/ACK approach, reliable transmission can be achieved at each hop. When there are more than one neighboring nodes receiving the broadcast transmission, we may use a round-robin approach, or a none-or-all approach. In a round-robin approach, the current node unicasts the packet to its neighbors in a one-by-one manner. Apparently this approach doesn't take advantage of the broadcast capability of wireless signal, and thus incurs high bandwidth consumption. In a none-or-all approach, after sending out the RTS message, the current node will wait for all neighboring nodes' CTS messages before it finally sends out the data packet, or it will abort this attempt of transmission and backoff and then retry again. Considering the potential number of neighboring nodes, this approach may incur much longer delay by the time all neighboring nodes are ready to receive data. The alternative approach without the reliability support from MAC layer is through redundancy. Allowing an appropriate degree of redundant retransmissions of duplicate packets, high degree of reachability is also obtainable.

A good comparison of various broadcasting techniques in MANETs can be found in [24], in which existing broadcast protocols have been categorized into four families: simple flooding, probability based methods, area based methods and neighbor knowledge based methods. A brief description of these categories follows.

- In a simple flooding scheme, each node in the network will forward the packet exactly one time. This process continues until all nodes in the network receive the packet. Simple flooding can be used as a simple protocol for broadcasting and multicasting in ad hoc networks with low node densities and/or high mobility.

- In a probabilistic scheme, intermediate nodes only rebroadcast with a certain probability. This approach is based on the understanding that in a dense network, nodal and network resources can be saved by having some nodes not rebroadcast the duplicate packets. A more refined probabilistic scheme is a counter-based approach in which upon receiving a broadcasted packet, the current node applies a random assessment delay (RAD) before it determines whether or not to rebroadcast the packet. If during this period the number of duplicate packets a node receives exceeds a given threshold, it assumes that its additional contribution is too marginal and decides not to rebroadcast the packet.

- In area based methods, intermediate nodes will evaluate additional coverage area based on all received duplicate packets. We can image that in a dense network there may be multiple nodes which are located very close to each other. In such situations, the majority of the coverage areas of these nodes overlaps each other. Based on estimated distance or location information, an intermediate node will determine whether or not to rebroadcast the received packet as illustrated in Figure 4.9.
- In neighborhood knowledge based methods, a node will determine whether or not to rebroadcast based on its neighbor list. Upon receiving a broadcasted packet, a node will check the previous node's neighbor list (which is included in the packet header). If it turns out that it would not reach any additional nodes, it will decide not to rebroadcast the packet.

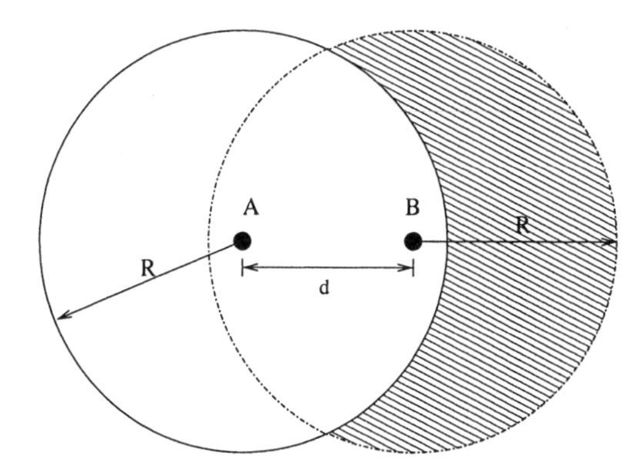

Figure 4.9. An example of area-based method: source node A sends a broadcast packet, and intermediate node B, based on its calculation of additional coverage area (shadowed in the figure), decides whether to rebroadcast the packet. Note that the additional coverage area of node B is a function of transmission radius R and nodal distance d. When d = R, the maximum additional coverage area is reached, which is about $0.61\pi R^2$.

A general framework on self-pruning-based broadcast redundancy reduction techniques in ad hoc networks was proposed in [25]. The authors proposed two neighborhood coverage conditions, which are used by intermediate nodes to determine whether or not to rebroadcast the packet upon receiving it. These coverage conditions depend on neighbor connectivity and history of visited nodes. Since global network information is costly, a distributed and local pruning process can be used to select the forwarding node set on the basis of local information such as k-hop neighborhood information. This selection process can be proactive (i.e., "up-to-date") or reactive (i.e., "on-the-fly"). Several existing proposals on broadcasting in ad hoc networks can be viewed as special cases of the coverage conditions with k-hop neighborhood information. Based on this proposed framework, the authors further propose new algorithms that combine features of previous works and show better performance.

4.5 Protocol Comparisons

It is difficult to make a quantitative, side-by-side comparison of all existing ad hoc multicast protocols due to the lack of such kind of performance evaluation results. In the context of ad hoc broadcasting, a good comparison can be found in [24]. For ad hoc multicasting, however, different research group have used different simulation environments and parameters, which greatly diminishes the comparability of the simulation result. We believe it is more practical to compare protocols qualitatively in the sense that which protocol(s) benefit most in what kind(s) of application scenarios. In other words, we believe it is difficult to make distinct statement that which multicast protocol is the best in existing literature due to the dynamic nature of ad hoc networks and the diversity of potential applications. It is more reasonable to believe that each of the protocols may fit in only some, not all, specific application scenarios.

In this section, we mainly consider three factors, namely, network size, network mobility, and multicast group size. We discuss how each of these factors may affect the performance of an ad hoc multicast protocol, and which class of existing protocols are most suited in these conditions.

4.5.1 Network Size

Due to different application scenarios, network size may vary in a vast range, from a small network with tens of nodes, to a large scale network with tens of thousands of nodes. Large scale network surely raises more challenges than a small network. If traffic locality is dominant, network size may not impose much difficulty in multicasting, and mesh-based protocols are more suited because group members are within proximity of one another and thus the overhead due to in-mesh flooding is reduced. However, in the cases where there is little traffic locality, multihop transmission is inevitable since most pairs of group members are multihop away.

In the case of large network size and small group size, if group members are scattered sparsely in the network area, stateless multicast and overlay multicast are more suited, while tree-based or mesh-based may incur too much overhead in route maintenance and redundant transmission. In the cases where both network size and group size are large, especially in cases where group members are clustered in several "hot spots", some form of hierarchy in constructing the forwarding infrastructure may help reduce routing and transmission overhead.

4.5.2 Network Mobility

Ad hoc networks may have different degrees of mobility. In a network with high degree of mobility, nodes move relatively fast, which results in rapidly changing topology. In a low mobility or static network, since nodes move slowly or remain stationary, the topology is relatively stable.

For a network with high mobility, mesh-based multicast protocols will outperform other multicasting methods. The path redundancy in mesh structure provides robustness against link breaks. For mesh-based ODMRP, the routing structure can be refreshed and fixed as a whole with one round of Join_Query and Join_Reply dialog. Thus, periodical dialogs can keep the routing mesh updated to the dynamic network topology. Moreover, with the path redundancy in the mesh structure, the update period can be larger than what is needed for tree-based protocols. Since a single link breakage will make the multicast tree disconnected, the overhead needed for maintaining the tree structure will be high. Overlay multicast and stateless multicast will also have performance degradations with high degree of mobility. For the overlay multicast protocol AMRoute, the overlay topology remains static under dynamic network topology. The optimality of the mulitcast tree will be significantly harmed under high mobility level. Problems such as congestion and buffer overflow will arise, and the protocol may fail to deliver a portion of the data packets [26] . The disadvantages of stateless multicast methods arise from its reliance on the underlying unicast protocol, which may have poorer performance under high mobility. For the stateless routing protocol DDM, intermediate nodes make forwarding decisions by querying the unicast protocol. It will cause high overhead for a unicast protocol to maintain a large set of routing entries. If the unicast routing table contains stale entries, the multicast forwarding will be compromised as well. For the same reasons, the performance of backbone-based protocols will be harmed by the higher degree of node mobility. On the other hand, if only some of the nodes are in highly mobile state, the multicast protocol can pick the slower and more stable nodes to form a relatively stable topology within the backbone.

There is a hybrid model which shows group mobility. Individual nodes form different groups based on their interest, and nodes in the same group move toward a common direction. In such a model, even though the group speed may be fast, the relative speed among group member is slow. M-LANMAR [10] attempts to exploit this model in facilitating team-based multicast in ad hoc networks.

4.5.3 Multicast Group Size

In addition to the network size, the multicast group size may be a more interesting factor affecting multicast performance. When multicast service scales up vertically (in terms of the group size) and horizontally (in terms of number of groups), how the protocol performance will be affected? The scalability issues of various protocols are discussed in [27].

When group size grows larger, tree-based protocols will incur high control overhead in maintaining a multicast tree of large size. When group member nodes are denser in the network, as the result of larger group, multicast meshes will achieve much higher forwarding efficiency. Backbone-based protocols are designed for achieving better scalability. The hierarchical method will take effect with larger groups. Stateless multicast protocols are designed for small group multicasting.

If there are multiple multicast groups and each group is of relatively small size, performance behavior will be different. Mesh-based protocols will not perform well. Forwarding efficiency will be much lower since the member nodes for each group are scarce in the network. Stateless multicast protocols are intrinsically suitable for this situation.

4.6 Overarching Issues

In this section, we provide an overview of the overarching issues that are important for almost all types of multicasting protocols. Specifically, we have addressed issues related to energy efficiency, reliability, quality of service, and security.

4.6.1 Energy Efficiency

Since mobile nodes in MANETs are typically driven by a limited battery source, it is essential to design energy conserving protocols for MANETs. Even in cases where energy is not a stringent source, reducing power consumption can result in less interference and better throughput over the wireless channels in MANETs. In this section we discuss various power-aware and/or energy efficient techniques for group communications in MANETs.

It is well known that wireless transmission contributes most part to the energy consumption in ad hoc networks. Naively, by reducing the number of nodes that participate in transmission, we can reduce the total energy for a broadcast/multicast process. To this end, many protocols we have described earlier share this approach of minimizing the forwarding node set. In [28], a passive clustering algorithm was proposed, which exploits data packets for cluster formation. Passive clustering is on-demand because it is executed only when there is user data traffic. Passive clustering can reduce node power consumption by eliminating the periodic, background control packet exchange.

Several proposed techniques on energy efficient broadcast/multicast share a common feature: combining minimum (or reduced) forward node set and power level selection. Broadcast Incremental Power (BIP) protocol was proposed in [29], which is a modified version of Prim's algorithm. Starting from the source node, BIP adds new nodes one at a time to the multicast tree. The decision on which node to add at each step is based on which node can be added with the minimum additional transmission energy. This new node may be reached from a leaf node, or from a parent node with increased transmission power. BIP is a greedy heuristic that requires global network information but may not generate the minimum cost tree. To address the problem of global network information as required in [29], [30] proposed a localized algorithm which requires only neighborhood information and attempts to take advantage of the broadcast nature of wireless transmission.

Energy consumption due to retransmission at data link layer when computing the minimum cost (energy) tree should be also considered while designing the protocols. Although quite a few efforts have been made in designing energy-efficient broadcast/multicast protocols, some other issues, such as how to address energy efficiency in presence of high mobility, how to factor in traffic condition in cases where contention-based MAC protocols are used, are still wide open topics.

4.6.2 Reliable Multicasting

Reliable group communication is a challenging task due to the dynamic nature in MANETs. When the node mobility is very high, flooding is a viable approach for reliable group communications in MANETs. When the mobility is too high, even simple flooding is insufficient for reliable multicast/broadcast in MANETs. In the following discussion, we assume that mobility is not so high that flooding or even more persistent variations of flooding become the only choice for reliable multicast/broadcast. In other words, with a range of nodal mobility and network dynamics, it is possible to search for more efficient and flexible alternatives for reliable group communication in MANETs.

The reliable broadcast protocol proposed in [31] is based on a clustering technique. It assumed that an underlying clustering protocol is in charge of constructing a clustered architecture covering the entire population of network hosts. A forwarding tree consisting of clusterheads is formed by distributing the broadcasted packets. Data packets are delivered to destination nodes via the cluster structure and acknowledgments travel backward along the path to the source node to achieve reliability. The reliable broadcast service here ensures that all destination nodes in the network deliver the same set of messages to the upper layer. This protocol gains reliability by paying the cost of maintaining the cluster structure pro-actively (i.e. even in the absence of traffic). Its efficiency also relies on the accuracy of forwarding tree and the underlying cluster structure, which is a challenging task when nodes move fast. To reduce the cost of maintaining the required structure information and improve the efficiency of broadcasting, the protocol in [32] was built on top of a low-cost unreliable broadcast operation. In this protocol, only a loose tree structure is maintained. To be specific, a node does not need to know which nodes are its children. Instead, a child node only needs to keep track of some possible parents. In the phase of distributing data packets, a counter-based unreliable broadcast scheme is used to reduce the overhead. During the phase of collecting acknowledgments, destination nodes send ACK to their parents, where and ACKs are combined and send to the source node. If the reversed tree structure is broken , neighboring nodes may exchange broadcasting history via a handshake procedure. When the source node is confirmed that all nodes have received the packet, it will send out a PURGE packet to notify all nodes to tear down the related data structure.

Hard guarantee of reliability in MANETs is not possible as the network size and mobility increase. A practical specification of probabilistic reliability was adopted in the RDG [22] protocol described earlier. RDG achieves a high level of reliability without relying on any inherent multicast primitive. In RDG, each group member only has a random partial view on the group, which result from the randomness of routing information that each node may have. RDG uses a pure gossip scheme in the sense that it gossips uniformly about multicast packets, negative acknowledgments, and membership information. The spread of the information is propelled mainly by a gossiper-push (each group member forwards multicast packet to a random subset of the group) but complemented by a gossiper-pull (multicast packets piggyback negative acknowledgments of the forwarding group member). Due to the non-deterministic characteristics of MANETs, the notion of probabilistic reliability seems quite fitting in this dynamic environment.

There has been a few efforts on MAC support for reliable group communications in MANETs. In [33], a new wireless ad hoc MAC protocol, Broadcast Medium Window (BMW) was proposed, the basic idea behind which is to reliable transmit each packet to each neighbor in a round robin fashion. BMW borrows concepts from IEEE 802.11, and attempts to achieve reliable broadcast support at MAC layer when traffic load is not too high. In circumstances where reliable transmission is counterproductive, BMW will reverts back to unreliable delivery of IEEE 802.11. The round robin approach in BMW does not take advantage of the broadcast nature of wireless signal, so it may incur much overhead by unicasting packets to each neighbor. Broadcast Support Multiple Access (BSMA), proposed in [34], incorporates the collision avoidance and RTS/CTS control frames of IEEE 802.11, and relies on negative acknowledgments (NACKs) to deliver broadcast packets reliably. Since the broadcast source node will wait for all destination neighbors to reply CTS, it may incur unnecessarily long delay before the data packet can successfully be transmitted.

4.6.3 QoS-Aware Multicasting

QoS is usually defined as a set of service requirements that needs to be met by the network while transporting a packet stream from a source to its destination. The network is expected to guarantee a set of measurable prespecified service attributes to the users in terms of end-to-end performance, such as delay, bandwidth, probability of packet loss, delay variance (jitter), etc. Power consumption and service coverage area are two other QoS attributes that are more specific to MANETs. With the increase in quality of service (QoS) needs in evolving applications, it is also desirable to support QoS-aware group communications in MANETs. The resource limitations and variability further add to the need for QoS provisioning in such networks. However, the characteristics of these networks make the QoS support a very complex process. Providing QoS in such a dynamic environment is very difficult. A number of works have been reported on topics of QoS provisioning in ad hoc networks [35]. QoS-aware group communication in ad hoc networks is still a very open problem.

Two compromising principles for QoS provisioning in the MANETs are: soft QoS and QoS adaptations.Soft QoS means that after the connection setup, there may exist transient periods of time when the QoS specification is not honored. However, we can quantify the level of QoS satisfaction by the fraction of total disruption time over the total connection time. This ratio should not be higher than a threshold. In a dynamic QoS approach, we can allow a resource reservation request to specify a range of values, rather than a single point. As available resources change, the network can readjust allocations within the reservation range. Similarly, it is desirable for the applications to be able to adapt to this kind of re-allocations. A good example of this case is the layered real-time video, which requires a minimum bandwidth assurance and allows for enhanced level QoS when additional resources are available. The QoS adaptation can be also done at various layers. The physical layer should take care of changes in transmission quality, for example, by adaptively increasing or decreasing the transmission power. Similarly, the link layer should react to the changes in link error rate, including the use of automatic repeat-request (ARQ) technique. A more sophisticated technique involves adaptive error correction mechanism which will increase or decrease the amount of error correction coding in response to the changes in transmission quality or the desired QoS. As the link layer takes care of the variable bit error rate, the main effect observed by the network layer will be a change in effective throughput (bandwidth) and delay.

4.6.4 Secure Multicasting

Security is an essential requirement in MANET environment. Its importance is amplified in group communications because of the involvement of more number of nodes. However, research in this area is still in the very beginning stage. Multicast in MANETs shares the same security issues as Internet multicast: receiver/source access control, group key management, multicast fingerprinting and secure multicast routing. The open group membership model makes joining a lightweight operation which can be conducted at any node. Besides, any host can send traffic to any multicast group, which the network tries to deliver to all group members. These properties, though the very spirit of multicast service provisioning, make the multicast service prone to the theft of service and/or denial of service attacks. Receiver and source access control is thus needed. If there is need to ensure the service is restricted to an authorized group of hosts, group data encryption with group key management is further needed as well. To prevent attacks towards the routing process, secure mechanisms is a necessity to group communication protocols.

Ad hoc networks have created a lot more additional challenges for implementation of the required security services beyond those that wireless communications encounter created when compared to wireline networks. The wireless broadcast media is more prone to both passive and active attacks. The MAC layer solutions to group key management and source authentication proposed for wireline networks need to be modified/enhanced to be able to adopted to wireless environment. Compared to other wireless communications such as cellular networks, ad hoc networks require even more sophisticated, efficient, and light-weighted security mechanisms in order to achieve the same security goals. These extra challenges are caused by, again, the dynamic characteristics of MANETs. First, MANETs lack trusted centralized infrastructure, which is often required in previous security proposals for wireline networks. Threshold-based and/or quorum-based approaches have been investigated toward addressing this problem. Second, the wireless links between nodes in a MANET is formed and torn down in an ad hoc fashion, which results in ephemeral relationships between nodes. The ephemeral relationships make it more difficult to build trust based on direct reciprocity. Third, proposed ad hoc multicast routing schemes are quite different from those for wireline networks. Additionally, MANET multicast has new group models such as geocasting. For some applications, especially in hostile environments such as battlefield communications, individual mobile nodes can be captured and compromised. As a result, the entire ad hoc network may be severely threatened. Finally, as always, for any solutions proposed for MANETs, overhead is of key concern due to the stringent nodal budgets (e.g. limited battery, slow processors, etc.) in many applications. Strong security mechanisms are needed, yet, the solutions need to be light-weighted no matter in terms of message overhead or computational cost. All these factors contribute to the difficulty in implementing security in group communication models of MANETs.

4.7 Conclusions and Future Directions

Several potential applications of mobile ad hoc networks have the need for point-to-multipoint communication. It is thus essential to provide multicasting support in ad hoc networks. In this chapter, we have presented a classification of multicasting protocols on the basis of their reliance on various types of nodes or networking layers. Description of the protocols including their salient characteristics and performance have been detailed in this chapter. Protocols for broadcasting techniques have been also presented. Several interesting overarching issues that are common to all protocol have been also analyzed.

With the advances in wireless technology and the applications of ad hoc networks, efficient multicasting support will become very critical. Future effort in this context should be targeted to energy efficient multicasting, QoS-aware multicasting, and cross-layer support for multicasting.

Acknowledgment

This work was supported in part by the National Science Foundation under the grants CCR-0296070 and ANI-0296034, and a generous gift from Hewlett Packard Corporation.

References

[1] S.E. Deering. Multicast routing in a datagram internetwork. Ph.D. thesis, Stanford University, Dec. 1991.

[2] P. Sinha, R. Sivakumar, and V. Bharghavan. MCEDAR: multicast core extraction distributed ad hoc routing. IEEE Wireless Commun. and Net. Conf. (WCNC), Sept. 1999, pp. 1313–17.

[3] J. J. Garcia-Luna-Aceves, E. L. Mada-uga. The core assisted mesh protocol. IEEE JSAC, Vol 17, No. 8, August 1999, pp.1380-1394.

[4] S.-J. Lee, M. Gerla, and C.-C. Chiang. On-demand multicast routing protocol(ODMRP). Proc. of IEEE WCNC'99, Sep. 1999.

[5] S. Lee, and C. Kim. Neighbor support ad hoc multicast routing protocol. In Proc. the First Annual Workshop on Mobile and Ad hoc Networking and Computing, 2000, pp. 37-44.

[6] B. Das, R. Sivakumar, and V. Bharghavan. Routing in ad hoc networks using a virtual backbone. In Proc. IEEE IC3N, 1997.

[7] J. Wu, and H. Li. A dominating set based routing scheme in ad hoc wireless networks. Proc. of the Third Int'l Workshop on Discrete Algorithms and Methods for Mobile Computing and Communications (DIAL M), pp. 7-14, Aug. 1999.

[8] K. Chen and K. Nahrstedt. Effective location-guided tree construction algorithms for small group multicast in MANET. In Proc. of INFOCOM, 2002, pp. 1180–89.

[9] L. Ji, and M. S. Corson. Differential destination multicast - A MANET multicast routing protocol for small groups. In Proc. of IEEE INFOCOM 2001: 1192-1202.

[10] Y. Yi, M. Gerla, and K. Obraczka. Scalable team multicast in wireless ad hoc networks exploiting coordinated motion. To appear in the Elsevier Ad Hoc Networks Journal.

[11] J. Xie, R. Talpade, T. McAuley, and M. Liu. AMRoute: Ad hoc multicast routing Protocol. ACM Mobile Networks and Applications (MONET) Journal, 7(6): 429-439, Dec 2002.

[12] C. Gui, and P. Mohapatra. Efficient overlay multicast for mobile ad hoc networks. IEEE WCNC 2003.

[13] C. E. Perkins and E. M. Royer. Ad hoc On-Demand Distance Vector Routing. Ad Hoc Networking, edited by C. E. Perkins, Addison-Wesley, 2001.

[14] E. M. Royer and C. E. Perkins. Multicast operation of the ad hoc on demand distance vector routing protocol. In Proc. of ACM MOBICOM, Aug. 1999, pp. 207–18.

[15] S.-J. Lee, W. Su, and M. Gerla. On-demand multicast routing protocol (ODMRP) for ad hoc networks. Internet-Draft, draft-ietf-manet-odmrp-02.txt, January 2000.

[16] C. Jaikaeo, and C.-C. Shen. Adaptive backbone-based multicast for ad hoc networks. In Proceedings of IEEE International Conference on Communications (ICC 2002), New York City, NY, April 28–May 2 2002.

[17] M. Gerla, C-C. Chiang, and L. Zhang. Tree multicast strategies in mobile, multihop wireless networks. ACM/Baltzer Journal of Mobile Networks and Applications (MONET), 1999.

[18] H. Eriksson. Mbone: The multicast backbone. CACM 37(8): 54-60 (1994)

[19] Y.H. Chu, S. G. Rao, H. Zhang. A case for end system multicast. SIGMETRICS 2000: 1-12.

[20] M. Kwon, S. Fahmy. Topology-aware overlay networks for group communication. NOSSDAV 2002: 127-136.

[21] R. Chandra, V. Ramasubramanian, and K. Birman. Anonymous gossip: Improving multicast reliability in mobile ad-hoc networks. In Proc. 21st International Conference on Distributed Computing Systems (ICDCS), pages 275–283, 2001.

[22] J. Luo, P.Th. Eugster, and J.-P. Hubaux. Route driven gossip: Probabilistic reliable multicast in ad hoc networks. In Proc. of INFOCOM'03, 2003.

[23] D.B. Johnson, D.A. Maltz, and J. Broch. DSR: The Dynamic Source Routing Protocol for Multi-Hop Wireless Ad Hoc Networks. Ad Hoc Networking, edited by C. E. Perkins, Addison-Wesley, 2001.

[24] B. Williams and T. Camp. Comparison of broadcasting techniques for mobile ad hoc networks. In Proc. of ACM Mobihoc '02, pp.194-205, 2002.

[25] J. Wu and F. Dai. Broadcasting in ad Hoc networks based on self-pruning. Proc. of the 22nd Annual Joint Conf. of IEEE Communication and Computer Society (INFOCOM), March 2003.

[26] S. J. Lee, W. Su, J. Hsu, M. Gerla, and R. Bagrodia. A performance comparison study of ad hoc wireless multicast protocols. In Proc. of IEEE INFOCOM 2000: 565-574.

[27] C. Gui and P. Mohapatra. Scalable multicasting in mobile ad hoc networks. IEEE Infocom 2004, Hong Kong, China, March 2004.

[28] Y. Yi, M. Gerla, and T. J. Kwon. Efficient flooding in ad hoc networks using on-demand (passive) cluster formation. In Proc. of ACM MobiHoc 2002.

[29] J. E. Wieselthier, G. D. Nguyen, and A. Ephremides. On the construction of energy-efficient broadcast and multicast trees in wireless networks. In Proc. of IEEE InfoCom 2000.

[30] J. Cartigny, D. Simplot, and I. Stojmenovic. Localized minimum-energy broadcasting in ad-hoc networks. In Proc. of IEEE InfoCom 2003.

[31] E. Pagani and G. P. Rossi. Providing reliable and fault tolerant broadcast delivery in mobile ad-hoc networks. Mobile Networks and Applications, vol. 4, pp. 175-192, 1999.

[32] C.S. Hsu and Y.C. Tseng. An efficient reliable broadcasting protocol for ad hoc networks. IASTED Networks, Parallel and Distributed Processing, and Applications (NPDPA), 2002, Japan, pp. 93-98.

[33] K. Tang, and M. Gerla. MAC reliable broadcast in ad hoc networks. In Proc. of MilCom 2001.

[34] K. Tang and M. Gerla. Random access MAC for efficient broadcast support in ad hoc networks. In Proc. of IEEE WCNC 2000.

[35] P. Mohapatra, J. Li, and C. Gui. QoS in mobile ad hoc networks. Special Issue on Next-Generation Wireless Multimedia Communications Systems in IEEE Wireless Communications Magazine, June 2003.

Chapter 5

TRANSPORT LAYER PROTOCOLS IN AD HOC NETWORKS

Karthikeyan Sundaresan
Georgia Institute of Technology
Atlanta, Georgia
sk@ece.gatech.edu

Seung-Jong Park
Georgia Institute of Technology
Atlanta, Georgia
sjpark@ece.gatech.edu

Raghupathy Sivakumar
Georgia Institute of Technology
Atlanta, Georgia
siva@ece.gatech.edu

Abstract The Transmission Control Protocol (TCP) is by far the most dominant transport protocol in the Internet and is the protocol of choice for most network applications. The focus of this chapter is to present approaches for ad-hoc networks that provide the same end-to-end semantics as TCP. In this regard, we first investigate the different problems experienced by TCP in ad-hoc networks, and provide insights into how the different design components of TCP relate to the characteristics of such networks. We then identify three major classes of approaches to improve the transport layer performance in ad-hoc networks. We present a protocol instance for each of the three approaches in detail and highlight the specific problems it addresses. We also discuss the trade-offs stemming from the adoption of each of the protocols considered.

Keywords: Ad-hoc networks, TCP

5.1 Introduction

The Open Systems Interconnection (OSI) reference model's fourth layer is the transport layer, which is responsible for reliable end-to-end communication and flow control functionalities. The TCP/IP protocol suite consists of the Transmission Control Protocol (TCP) and User Datagram Protocol (UDP) as the transport protocols[1]. UDP is a simplistic transport layer solution that merely provides labeling functionality for applications. Any further functionality in terms of reliability, flow-control, etc., is pushed up into the application. Good examples of applications that rely on UDP are multimedia applications.

In contrast, TCP is a complex transport layer protocol that provides applications with *reliable*, *end-to-end*, *connection-oriented*, and *in-sequence* data delivery. It performs both flow control, and congestion control on behalf of the applications, recovers from packet losses in the network, and handles re-sequencing of data at the receiver. Of the traffic carried by the Internet, TCP accounts for about 90% of the bytes, with UDP accounting for the most of the remaining traffic [1]. Although the use of UDP is increasing due to the increase in the usage of multimedia applications, TCP continues to play a dominant role in the Internet.

In this chapter, we focus on the design of transport layer protocols for ad-hoc networks. The unique characteristics of ad-hoc network environments clearly will impact the requirements imposed on a transport layer protocol. Thus, it is interesting to both investigate from a top-down standpoint how a protocol as well established as TCP would work over such environments, and to study from a bottom-up standpoint what kind of transport layer behavior ad-hoc networks necessitate. Given the dominance of TCP in terms of being the protocol of choice for network applications, we restrict the focus of the chapter to protocols that can support the same end-to-end semantics of reliable, in-sequence, data-delivery as TCP.

We first present detailed arguments on how each of TCP's design elements relate to the characteristics of ad-hoc networks, and motivate whether or not a fundamental re-design of the transport layer protocol is even necessary. We arrive at the conclusion that such a re-design is indeed necessary. However, we also identify the fact that issues of backward compatibility in certain environments (e.g. mobile host communicating with a static Internet server through an ad-hoc network) might require staying within the TCP paradigm. In such cases, the focus should then be on approaches to improve performance given that TCP or a TCP-based protocol is used.

Thus, we discuss three broad classes of approaches to improve transport layer performance over ad-hoc networks:

---

[1]The TCP/IP protocol suite consists of four layers with the transport protocols at the third layer.

- *Modified TCP:* This represents a class of transport layer approaches, where minor modifications are made to the TCP protocol to adapt it to the characteristics of an ad-hoc network, but the fundamental elements of TCP are still retained [2, 3].

- *TCP aware Cross Layer Solutions:* This represents a class of lower layer approaches that *hide* from TCP the unique characteristics of ad-hoc networks, and thus necessitate minimal changes to TCP. Such approaches can be used in tandem with the approaches in the previous class.

- *Ad-hoc Transport Protocols:* Finally, this represents a class of *new* built-from-scratch transport protocols that are built specifically for the characteristics of an ad-hoc network, and are not necessarily TCP-like.

For each of the classes of approaches, we discuss one representative protocol, investigate its mechanisms, and highlight its performance. We also provide discussions on trade-offs between the different classes of approaches, wherever applicable.

The rest of the chapter is organized as follows: Section 5.2 consists of a brief overview of the TCP protocol, and an in-depth study of the appropriateness of the design elements of TCP for ad-hoc networks. Section 5.3 is a high level introduction to the three classes of approaches considered. Sections 5.4, 5.5 and 5.6 discuss in detail specific protocol instances of the different approaches. Finally, Section 5.7 summarizes the key conclusions of the discussions in the chapter.

5.2 TCP and Ad-hoc Networks

In this section, we investigate in detail whether or not TCP's fundamental design elements are appropriate for ad-hoc networks. Note that the performance of TCP over one-hop wireless cellular data networks is well studied in related works. Interested readers are referred to [4] for a detailed exposition on the issues involved in operating TCP over a cellular environment and the various approaches that have been proposed in literature toward the design of a transport protocol for the cellular environment. However, the characteristics of multi-hop wireless ad-hoc networks are significantly different from those of the cellular environment and hence this calls for a study of TCP's operation over ad-hoc networks as well.

In the rest of the section, we first outline the different components of a TCP connection, and then investigate how the components impact TCP's performance in ad-hoc networks.

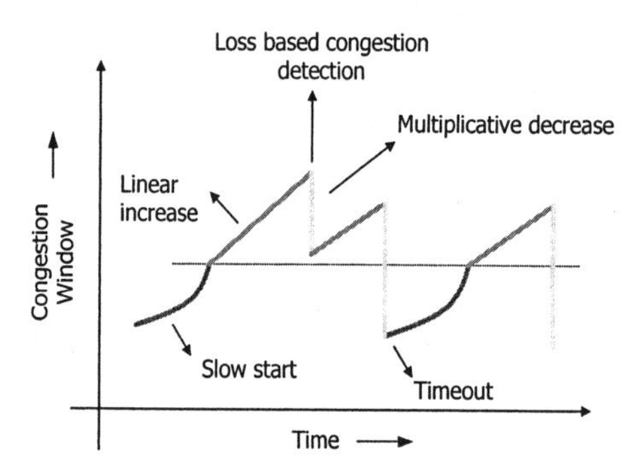

Figure 5.1. Number of route errors

5.2.1 TCP Background

We use the different phases in a TCP connection's congestion window progression to explain TCP's fundamental design elements. TCP uses *window-based transmissions*. The number of unacknowledged packets transmitted on the channel is determined by the size of the congestion window. Hence, the progression of the congestion window can be directly related to the throughput enjoyed by the connection. Further, the arrival of ACKs from the TCP receiver drives the progression of the sender's congestion window. Initially, when a connection is initiated, the TCP sender enters the *slow-start* phase. In this phase, the congestion window is increased by one for every ACK that is received. Hence, there is an exponential increase of the congestion window, with the window doubling every round-trip time. Once the window size exceeds an $ssthresh$ threshold, the window increases by one for every round-trip time (rtt). This phase is referred to as the *congestion avoidance* phase where the progression of window is linear. The sender continues to perform *linear increase*, probing for more available network bandwidth. The increase continues till a loss is perceived. On experiencing a loss, the sender infers congestion (*loss-based congestion detection*) and reduces the congestion window. The nature of reduction depends on the nature of loss. If the loss is notified by the arrival of triple duplicate ACKs, then a *multiplicative decrease* of the window is performed, wherein the window is decreased to half its current value, and the connection enters the congestion avoidance phase. On the other hand, if the loss is detected through a *retransmission timeout*, then the window is reduced to one and the connection enters the slow-start phase again. These basic elements in the anatomy of a TCP connection are illustrated in Figure 5.1.

In the rest of the section, we use arguments substantiated with some packet level network simulation results to highlight the appropriateness of the above mechanisms to the specific characteristics of ad-hoc networks. For all the simulations, FTP is used as the application generating traffic. The *Newreno* version of TCP is used with Dynamic Source Routing (DSR) as the routing protocol. Carrier Sense Multiple Access with Collision Avoidance (CSMA/CA) in the Distributed Co-ordination Function (DCF) mode is used as the medium access control protocol. The two ray ground reflection model is used as the propagation model with a cross-over distance of 100m. The cross-over distance denotes the radius within which the path loss coefficient is two and beyond which the path loss coefficient is four. Inside the cross-over distance a line of sight model is assumed.

5.2.2 *Window-based Transmissions*

One of the motivating factors for TCP being window based is the avoidance of the maintenance of any fine-grained transmission timers on a per-flow basis. Instead, TCP uses the principle of self-clocking (ACKs triggering further data transmissions) for connection progression. For wireline environments, where per-flow bandwidths can scale up to several megabits per second, such a design choice is clearly essential. However, the use of a window based transmission mechanism in ad-hoc networks may result in the critical problem of burstiness in packet transmissions.

Thus, if several ACKs arrive back-to-back at the sender, a burst of data packets will be transmitted by the sender even if it were in the congestion avoidance phase (where one packet will be transmitted for every incoming ACK). Unfortunately, *ACK bunching* or several ACKs arriving at the same time is a norm in ad-hoc networks because of the short-term unfairness of the CSMA/CA MAC protocol typically used in such networks. [5] provides a good exposition on the short term unfairness properties of CSMA/CA. Such short-term unfairness results in the data stream of a TCP connection assuming control of the channel for a short period, followed by the ACK stream assuming control of the channel for a short period. Interestingly, such a phenomenon will occur even when the ACK stream does not traverse the exact same path as the data stream. This is because even if the paths were completely disjoint, the vicinity (2-hop region in the case of CSMA/CA) of the TCP sender and the vicinity of the TCP receiver still are common contention areas for the data and ACK streams. Figure 5.2(a) shows the TCP sequence number progression (at the sender) in a single TCP connection scenario. It can be seen that the transmissions occur in periods of bursts and are interspersed with periods of inactivity due to the arrival of ACKs. The impact of such burstiness of traffic has two undesirable effects:

- *Varying round-trip time estimates:* TCP relies on an accurate round-trip time (*rtt*) calculation to appropriately set the timer for its retransmission

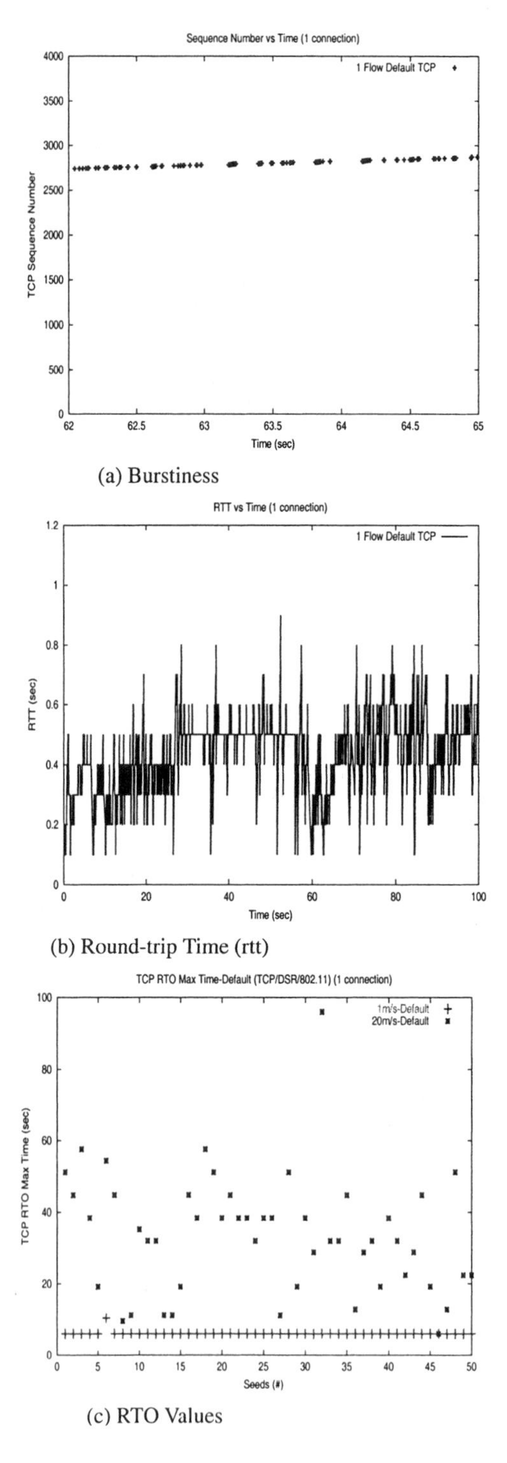

(a) Burstiness

(b) Round-trip Time (rtt)

(c) RTO Values

Figure 5.2. Round-trip Time and Timeouts (1 Flow)

timeout (RTO). Coupled with the low bandwidths available to flows, the burstiness results in artificially inflating the round-trip time estimates for packets later in a burst. Essentially, the round-trip time of a packet is impacted by the transmission delay of the previous packets in the burst due to the typically small available rates. TCP sets its RTO value to $rtt_{avg} + 4 * rtt_{dev}$, where rtt_{avg} is the exponential average of *rtt* samples observed, and rtt_{dev} is the standard deviation of the *rtt* samples. Hence, when *rtt* samples vary widely due to the burstiness, the RTO values are highly inflated, potentially resulting in significantly delayed loss recovery (and hence under-utilization). Figures 5.2(b) and (c) show the variation in *rtt* and the average maximum RTO values for the single connection, where it can be observed that RTO values increase with an increase in mobility.

- *Higher induced load:* Spatial re-use in an ad-hoc network is the capability of the network to support multiple spatially disjoint transmissions. Unfortunately, due to the burstiness and the short term capture of channel by either the data stream or the ACK stream, the load on the underlying channel can be higher than the average offered load. We refer to the artificially (short-term) increased load on the underlying channel as the induced load. If the offered load is not high, the higher induced load will not result in any major performance degradation. However, if the offered load itself is high (around the peak scalability of the underlying MAC layer's utilization curve), the utilization at the MAC layer can suffer significantly.

5.2.3 *Slow Start*

TCP performs slow start both during connection initiation, and after experiencing a retransmission timeout. For both cases, the goal of slow-start is to *probe* for the available bandwidth for the connection. When a connection is in the slow-start phase, TCP responds with two data packet transmissions for every incoming ACK. While this exacerbates the burstiness problem discussed earlier, there are two other problems associated with the slow-start mechanism in the context of ad-hoc networks:

- *Under-utilization of network resources:* Although slow-start uses an exponential increase of the congestion window size, the increase mechanism is still non-aggressive by design as it can take several *rtt* periods before a connection operates at its true available bandwidth. This is not a serious problem in wireline networks as connections are expected to spend most of their lifetimes in the congestion avoidance phase. However, because of the dynamic nature of ad-hoc networks, connections are prone

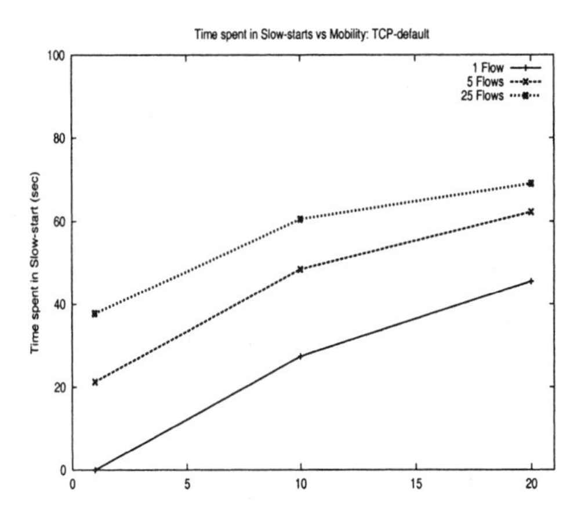

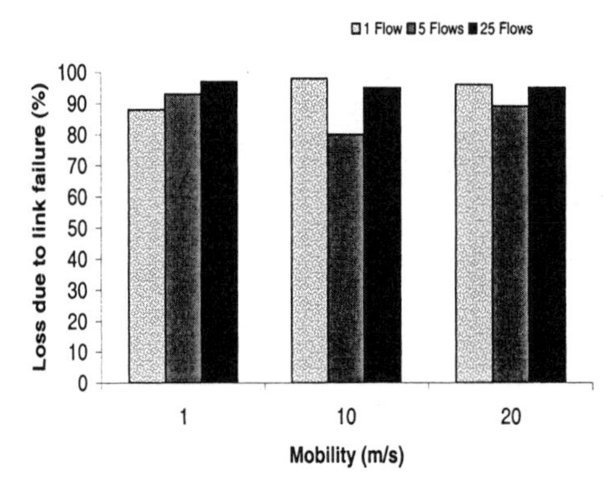

(b) % of losses due to route-failures

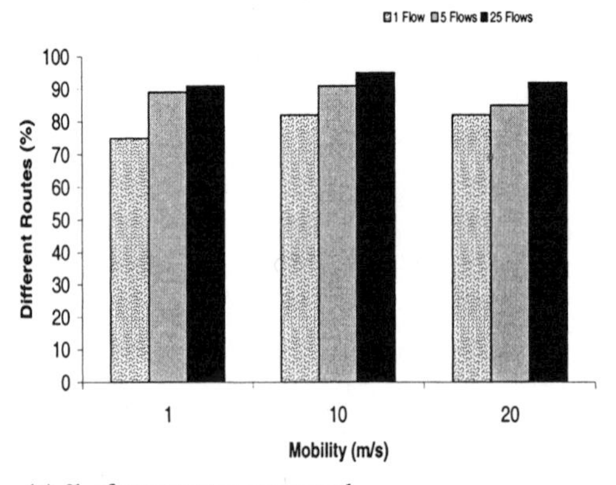

(c) % of new routes computed

Figure 5.3. Slow-start and Loss-based Congestion Detection

to frequent losses which in turn result in frequent timeouts and hence more slow-start phases. Figure 5.3(a) presents the average time spent in slow-start by the connections during the 100 second simulation. It can be observed that connections spend a considerable amount of time in the slow-start phase, with the proportion of time going above 50% for the higher loads. Essentially, this means that connections spend a significant portion of their lifetime probing for the available bandwidth in lieu of operating at the available bandwidth.

- *Unfairness:* TCP's fairness properties are firmly dependent upon the contending connections operating in congestion avoidance. When connections operate primarily in the slow-start phase, the fairness properties of TCP are more likely to be violated, since the slow start phase in TCP is not designed keeping the fairness properties in mind.

5.2.4 *Loss-based Congestion Indication*

TCP detects congestion through the occurrence of losses. While congestion is by far the main source of losses in wireline networks, it is well known that this is not the case in wireless networks. In conventional cellular wireless networks, non-negligible random wireless channel error rates also contribute to losses. In ad-hoc networks, in addition to congestion and random wireless errors, mobility serves as another primary contributor to losses perceived by connections. Random wireless errors are addressed to some extent through the use of a semi-reliable MAC layer such as CSMA/CA that uses a positive ACK after data reception to indicate successful reception of a packet. Interestingly, CSMA/CA does not distinguish between whether a link is down because of the other end moving out of range, or because of high contention at the receiver. In either case, after attempting to transmit to a receiver for a finite number of times, the MAC layer concludes a link failure and informs the higher layers accordingly. Most routing protocols designed for ad-hoc networks [29, 47] rely on such MAC feedback to trigger route-failure notification to the source.

Losses in ad-hoc networks can be classified into either link failure induced, or congestion induced (interface queue overflows), with most of the losses being due to link failures. Figure 5.3(b) presents the percentage of the number of losses due to route (link) failures for different rates of mobility and loads. It can be observed that in all the scenarios, more than 80% of the losses in the network are due to link failures. Note that a link failure can be inferred by the MAC layer even when it is not able to reach a neighbor due to severe congestion. However, irrespective of the true cause of link failure inference, the source will be notified of a route failure and a new route computation will be performed. Figure 5.3(c) shows the percentage of time when the old route is again chosen by the route computation mechanism. It can be observed that about 90% of the time, a different route is chosen. Essentially, most losses in ad-hoc networks occur as a result of route failures (in reality, the MAC and routing layer *perceive* most of the losses as due to route failures), and hence treating losses as an indication of congestion turns out to be inappropriate.

5.2.5 *Linear Increase Multiplicative Decrease*

Once the available bandwidth has been probed by the slow start mechanism, TCP enters the congestion avoidance phase where it decreases the rate of increase in the amount of data pumped into the network, so as not to cause congestion. Hence, in this phase the congestion window is increased only linearly. Congestion avoidance is also performed immediately after a multiplicative window decrease induced by the reception of a triple dulplicate ACK. The linear increase phase of TCP has the same drawback of slow-start – slow convergence to the optimal operating bandwidth, and hence vulnerability to route failures before the optimal bandwidth is attained.

The multiplicative decrease on the other hand is inappropriate for the reasons discussed in Section 5.2.4. Essentially, most loss events in an ad-hoc network are due to route failures, or are perceived to be due to route failures by the underlying layers. Hence, more often than not, a loss event experienced by a connection is followed up by a route change (see Figure 5.3(c)). While TCP's multiplicative decrease is an appropriate reaction to congestion, it is definitely not an appropriate action to take when a route change has occurred, especially given that most of the time a different route is chosen. Ideally, when a route change occurs, TCP should enter its bandwidth estimation phase as its old congestion window state is not relevant to the new route.

5.2.6 *Dependence on ACKs and Retransmission Timeouts*

The occurrence of packet losses are identified by the TCP sender by the arrival of triple duplicate ACKs and through retransmission timeouts. The ACK stream not only helps achieve the reliability functionality of the TCP protocol, but is also used to clock the transmission of data packets at the TCP sender. In short, TCP relies on the periodic arrival of ACKs both to ensure reliability and to perform effective congestion control. Most implementations of the TCP receiver send one ACK for every two packets received. This dependence on ACKs results in two problems for ad-hoc networks: (i) Due to the overhead (about 100 bytes) associated with the request-to-send (RTS), clear-to-send (CTS), and ACK packets used by the CSMA/CA protocol, TCP ACKs sent from the receiver to the sender can amount to 10-20% of the data stream rate. If the forward and reverse paths happen to be the same,[2] the ACK traffic in the reverse path

[2]Routing protocols in ad-hoc networks may or may not choose the same path in two directions.

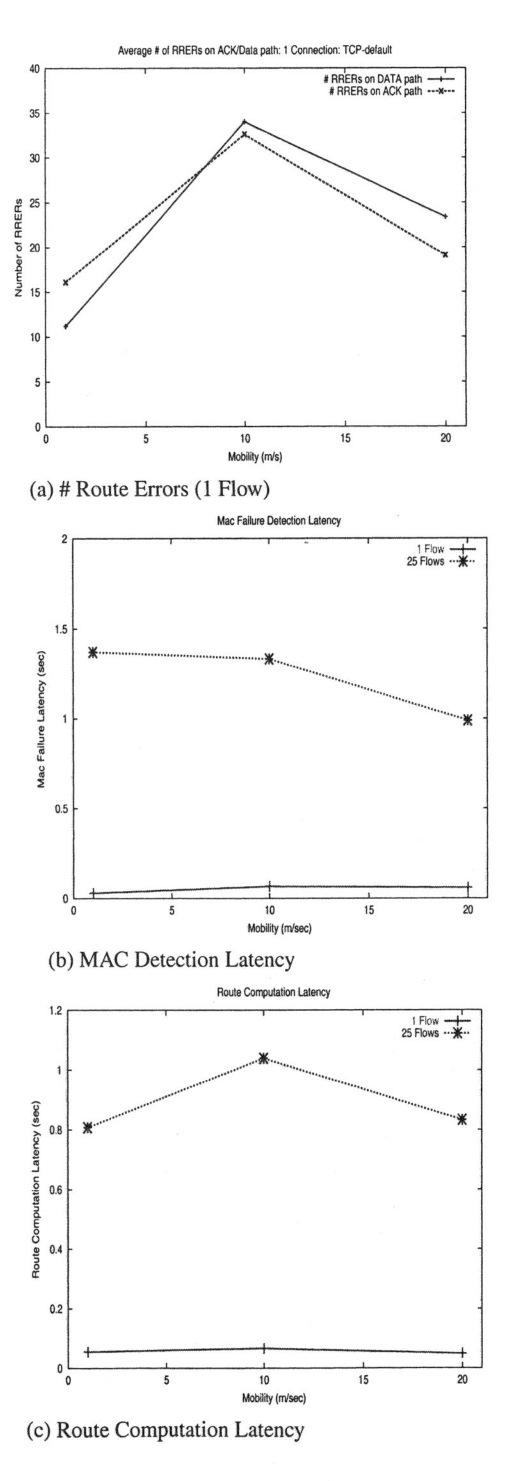

(a) # Route Errors (1 Flow)

(b) MAC Detection Latency

(c) Route Computation Latency

Figure 5.4. Route Errors and Impact of Losses

will contend with the data stream on the forward path and reduce the rate enjoyed by the data stream. (ii) If the forward and reverse paths are not the same, the progress of the TCP connection will be dependent on both the forward path and reverse path reliability. Thus, the chances of a connection stalling increase when different paths are used. Note that even if the forward and reverse paths are different, due to the shared channel in the vicinity of the sender and the vicinity of the receiver, the data and ACK streams will still contend with each other. Figure 5.4(a) shows the number of times the data stream and the ACK stream experience independent path failures for the 1 flow scenario. It can be observed that the forward and reverse paths experience the same order of magnitude of failures.

TCP relies on retransmission timeouts as a backup loss detection mechanism. As described in Section 5.2.2, the RTO value for a TCP connection can be considerably inflated and vastly different from the optimal value. Figure 5.2(c) presents the average of the maximum RTO values set by connections during their lifetimes. It can be observed that for higher rates of mobility, the maximum RTO values scale up to few tens of seconds. This is true even in the case of heavy loads. This can result in significant time delays in loss recovery, and hence result in gross under-utilization of the available bandwidth.

5.2.7 *Absolute Impact of Losses*

In the discussions thus far in the section, we have touched upon the negative impact of mobility related losses on TCP's performance. Losses, in addition to being inaccurate indicators of congestion for TCP, also have an absolute impact on the throughput performance of the TCP connection. In this section, we profile some of the directly contributing causes for such losses.

Impact of MAC Layer. The MAC layer is responsible for detecting the failure of a link due to congestion or mobility. Since the MAC layer (IEEE 802.11 DCF) has to go through the cycle of multiple retransmissions before concluding link failure, there is a distinct component associated with the time taken to actually detect link failure since the occurrence of the failure. Importantly, the detection time increases with increasing load in the network. A high MAC detection time will result in a higher likelihood of the TCP source pumping in more packets (upto a window's worth) into the broken path, with all the packets being lost and the source eventually experiencing a timeout.

When a link failure is detected by the MAC layer, the link failure indication (in DSR) is sent only to the source of the packet that triggered the detection. If another source is using the same link in the path to its destination, the node upstream of the link failure will wait till its MAC layer receives a packet from the other source. Then the MAC layer will go through its cycle of retransmissions to detect the link failure and only then would that source be informed of the link failure. This also contributes to the delay after which a source realizes that a path is broken, consequently increasing the probability of timeouts. Figure 5.4(b) shows the latency involved in the MAC layer detecting a link failure. It can be observed that for higher loads, the latency could be in the order of a few seconds.

Both of the above factors directly contribute to more number of losses occuring in the network, and thus impact the throughput performance of network connections.

Impact of Routing Layer. The characteristics of the underlying routing protocol have a significant impact on TCP's performance. Some of the important ones are outlined below.

In most of the reactive routing protocols (such as DSR), there is a provision for the routing layer at the upstream node of a broken link to send back a path failure message to the source. Once the source is informed of the path failure, it initiates a new route computation. Any packet originating at the source during this route-recomputation phase *does not have a route*. This directly increases the fraction of time that packets in the routing layer spend without a route to the destination during a connection's lifetime. Further, the time taken to recompute the new route also increases with increasing load. This can be observed in Figure 5.4(c) where the latency involved in route computation is presented.

In addition to the absolute impact of not having a route in the route computation phase, TCP is also likely to experience timeouts during each route computation time, especially in the heavy load scenario where route computation time is around a couple of seconds. Furthermore, successive timeouts and the resulting back-offs could potentially result in the stalling of the data connection.

Finally, it is in the best interest of the connection to minimize the number of route failures resulting from the routing protocol's operation. This is because, the number of route failures directly influence the above two factors. As the number of route errors increases, the fraction of time a packet spends without a route at the routing layer increases, consequently increasing the probability of the expiry of TCP's retransmission timer.

5.3 Transport Layer for Ad-hoc Networks: Overview

Existing approaches to improve transport layer performance over ad-hoc networks fall under three broad categories: (i) Modifying TCP to handle the characteristics of an ad-hoc network, (ii) Cross-layer TCP aware modifications to the lower layers of the protocol stack to hide from TCP the vagaries of an ad-hoc network, (iii) Built-from-scratch transport protocols that involve a fully re-designed transport layer approach suited for ad-hoc networks. In the rest of the section, we provide an overview of each of the above three approaches, and in the ensuing sections we elaborate on the details of the approaches.

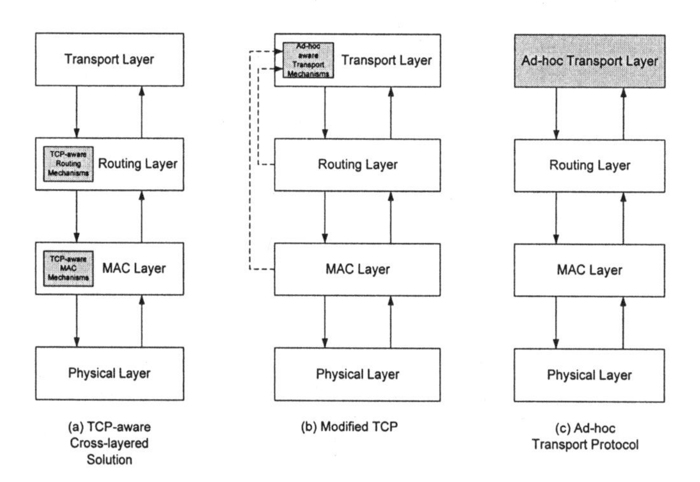

Figure 5.5. Classification of Approaches

- A straight-forward and simplistic approach is to retain TCP as the transport protocol, but make it mobility-aware by supplementing it with additional mechanisms, along with simple support from the lower layers to overcome the negative impacts of mobility. We refer to this approach as the *Modified TCP* approach. While the transport layer does require some changes in this approach, the level and complexity of changes can be viewed as a trade-off with the performance improvements possible. A key advantage of such an approach is that the general behavior of the protocol is similar to that of TCP, and hence backward compatibility issues, when mobile-hosts talk to static-hosts in the Internet, do not arise. However, an obvious drawback is that the problems identified with the design elements of TCP in Section 5.2 are still left un-addressed.

- In the second approach, TCP is hidden from the underlying network characteristics through appropriately designed lower layer protocols. Hence, all the required mechanisms to mask out the negative effects of mobility on TCP, are implemented at the MAC and routing layers. This requires no changes to TCP's operation. This approach is in fact more suitable for addressing backward compatibility issues raised earlier. We refer to this approach as *TCP-aware Cross-layered Solutions*. Note that unlike in the first approach, where the underlying protocols are to a large extent TCP unaware, this approach requires the lower layers to possess a close awareness of TCP's properties and behavior. Also, as in the first approach, since the mechanisms are all implemented at the routing and

MAC layers without changing TCP, some of the inherent problems resulting from TCP's key design elements cannot be addressed.

- Finally, the third approach is to consider a transport layer design that is drawn from scratch, and tailored specifically for the characteristics of an ad-hoc network. We refer to this approach as the *Ad-hoc Transport Protocol* approach. An obvious drawback of such an approach is that hosts in the ad-hoc network will now possess a transport layer protocol that is different from TCP. While this is not a problem in stand-alone dedicated ad-hoc networks such as those in military applications, it is an issue when ad-hoc networks are seen to "hang-off" from the Internet. As we show later in the chapter, such an approach can provide connections with the best possible performance. Such an approach can also be used as a *bench-mark* for the earlier two approaches.

Figure 5.5 illustrates the layers that exhibit changes in the protocol stack for the three classes of approaches. In the following sections, we present one instance of each of the above classes of approaches in detail. We also discuss other related work under the three approach categories.

5.4 Modified TCP

The first approach involves effecting changes to the TCP protocol with support from the underlying layers in order to mask out the problems arising from mobility. A protocol that falls in this class is the *ELFN (Explicit Link Failure Notification)* protocol proposed by Holland et al. [2]. ELFN employs simple support from the network and lower layers to achieve the purpose. The bulk of the mechanisms in ELFN reside in the transport layer, and can be viewed as being additional to those already present in TCP. The objective of ELFN is to provide the TCP sender with information about link and route failures so that it can avoid responding to the failures as if congestion occurred, and consequently reduce any unnecessary degradation in performance. In the rest of the paper we refer to this approach as simply TCP-ELFN.

Mechanisms

The salient features of TCP-ELFN are:

- When a link failure occurs, the node upstream of the failure link sends back an ELFN message to the source of every TCP connection using that link. A link failure is said to occur when the MAC layer is unable to successfully deliver a packet across the link after trying for a threshold number of times. The notification message is sent by modifying DSR's route failure message to carry a payload similar to the "host unreachable" ICMP message.

- When the routing protocol of a source node receives the link failure message, it first sends a route re-computation request. In addition, it sends this ELFN message to the transport layer. When the TCP source receives the ELFN message, it *freezes* its state that includes entities such as congestion window, retransmission timers, and enters a "stand-by" mode. While on stand-by, a packet is sent at periodic intervals to probe the network to see if a route has been established. If an ACK is received, then it leaves the stand-by mode, restores its retransmission timers, and continues as normal. Packet probing instead of an explicit notice is used to signal that a route has been re-established. When the connection enters the "stand-by" mode, an associated timer is started. If the timer expires before a new route is computed, then the connection is made to come out of the "stand-by" mode to experience losses, and enter slow-start with an initial congestion window size of one.

- Variations in some parameters and actions of the protocol are also possible. In particular, the following variations have been studied: (a) Variations in the length of the interval between probe packets, (b) Modifications to the RTO and congestion window upon restoration of the route, and (c) Different choices for which packet to send as a probe. The impact of each of these variations is discussed in [2].

Performance

The key advantage of ELFN is that it hides the latency of route re-computations upon path failures from the TCP layer, and prevents TCP from reacting adversely to route-failures. Specifically, the benefits are obtained from the following factors: (i) Freezing TCP's state prevents it from cutting down its window to one and entering the slow start phase. This in turn reduces the number of retransmission timeouts experienced by the connection. As observed in Figure 5.6(a), for the one flow case, the timeout values are not as scattered as they are in the default TCP case for higher mobility rates in Figure 5.2(c), and (ii) Stopping further transmissions till a new route has been computed, thereby preventing those packets from being lost along the broken route. This directly reduces the number of losses suffered by the connection and consequently increases the throughput as observed in Figures 5.6(b) and (c).

Trade-offs

TCP-ELFN provides the flexibility to retain TCP's existing components, while masking any mobility related problems through additional transport layer

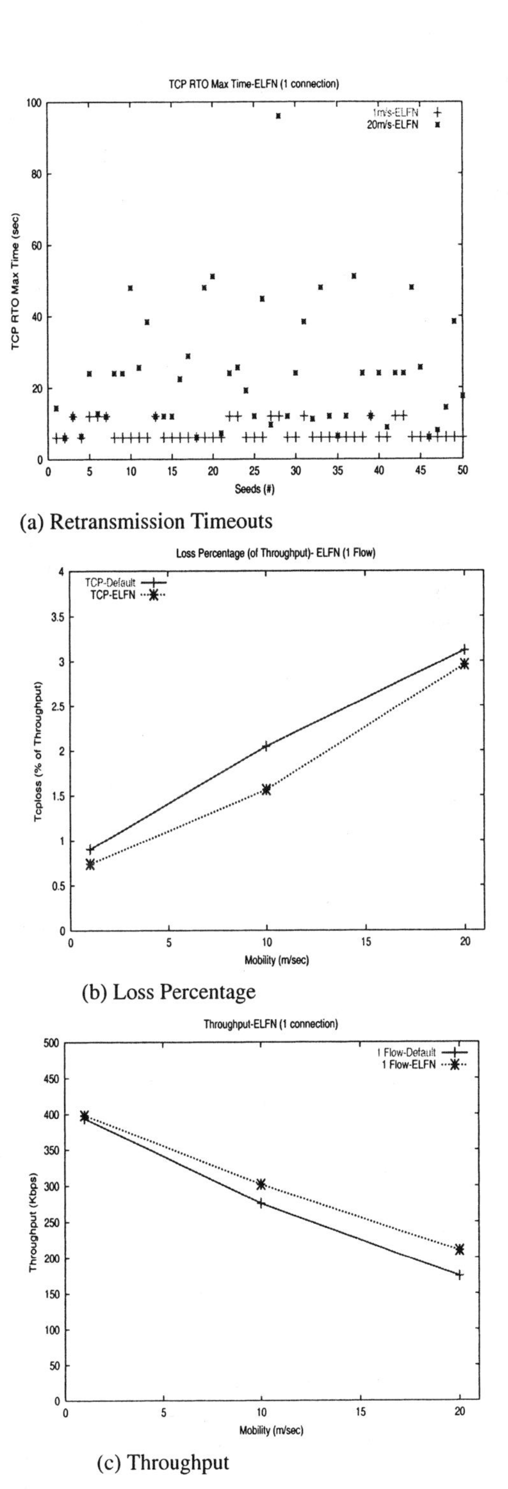

(a) Retransmission Timeouts

(b) Loss Percentage

(c) Throughput

Figure 5.6. TCP-ELFN (1 Flow)

mechanisms. However, a drawback with TCP-ELFN is that the connections could suffer from lower throughput in heavily loaded static scenarios [8]. This is because the MAC layer is incapable of determining if a loss is due to route failure, or simply due to high contention. Hence, in heavily loaded static scenarios, ELFN messages would be generated even in the absence of route failures. This would make the connection enter the "stand-by" mode and wait for a new route computation to restore its state, thereby causing a degradation in throughput. The choice of the probing interval is a critical parameter. While a large value could increase the time for computation of a route, thereby reducing the throughput, a small value could increase the contention in the network due to the frequent probe packets that are generated. Furthermore, ELFN also tries to mask out only the negative impacts arising from route failures caused due to mobility by freezing TCP's state during new route computations. However, some of the basic characteristics of TCP that degrade its performance in ad-hoc networks are still left un-addressed.

Related Work

There have been some recent works [8] [9] [10] [11] that discuss the effect of mobility on TCP performance and suggest various transport layer mechanisms to solve the problems caused due to mobility. [8] studies the performance of ELFN on static and dynamic networks and corroborates the results obtained in [2]. [9] discusses a mechanism called TCP-Feedback, which uses route failure and re-establishment notifications to provide feedback to TCP, and thus reduce the number of packet re-transmissions and TCP back- offs during route calculation, to improve throughput. However, this mechanism is evaluated in a simple one-hop wireless network. [10] studies the performance of TCP on three different routing protocols and proposes a heuristic called fixed RTO, which essentially freezes the TCP RTO value whenever there is a route loss. It also evaluates the effectiveness of TCP's selective and delayed acknowledgments in improving the performance. [11] provides a transport layer solution to improving TCP performance. It introduces a thin layer between the transport and underlying routing layers, which puts TCP into persist mode whenever the network gets disconnected or there are packet losses due to high bit error rate. Thus, this thin layer acts as a shield to TCP, protecting it from the underlying behavior of an ad-hoc network.

5.5 TCP-aware Cross-layered Solutions

An approach that falls under this category is the *Atra* framework proposed by Anantharaman et. al. [12]. The framework comprises of mechanisms at both the routing and medium access control layers and does not necessitate any changes to TCP. The mechanisms in the *Atra* framework are based on the following three goals: (i) To minimize the probability of route failures; (ii) To predict route failures in advance and thus enable the source to recompute an alternate route before the existing route fails; and (iii) To minimize the latency in conveying route failure information to the source, for route failures that are not successfully predicted. The *Atra* framework consists of three mechanisms targeted toward each of the above goals respectively:

Mechanisms

The key mechanisms in Atra are the following:

- *Symmetric Route Pinning:*

  The DSR routing protocol does not explicitly use symmetric routes between a source and a destination, i.e. the route taken from the source to the destination can be different from the route taken from the destination to the source. While the use of asymmetric routes is not an issue in a static network, in a dynamic network where nodes are mobile using an asymmetric path, increases the probability of route failure for a connection.

  Specifically, a TCP connection will stall irrespective of whether the forward path is broken or the reverse path is broken. Taking the simple scenario of using two edge-disjoint routes for the data and the ACK paths with hop lengths of $h1$ and $h2$ respectively, and assuming a uniform probability of link failure p for all links in the network (which is not unrealistic given the use of the random way-point mobility model), the probability of a path failure is $1-(1-p)^{(h1+h2)}$. Hence, in the first mechanism within the *Atra* framework called *symmetric route pinning* (SRP), the ACK path of a TCP connection is always kept the same as the data path in order to reduce the probability of route failures. The mechanism implemented at the DSR layer does the route pinning only for uni-directional communication. The reasoning is as follows: while it is true that the forward path progression can be asynchronous to the reverse path progression, performing route pinning to piggybacked ACKs in bi-directional communication can severely increase the congestion along the path whereas in the case of asymmetric paths, implicit load balancing is performed.

- *Route Failure Prediction:*

  The symmetric route pinning mechanism merely reduces the probability of route failures for a connection. Hence, the second mechanism in *Atra* attempts to predict the occurrence of a link failure by monitoring the signal strength of the packets received from the corresponding neighbor. Based on the progression of signal strengths of packet receptions from a particular neighbor, a node predicts the occurrence of the link failure. Maintaining a history of the progression enables nodes to dynamically profile the speed at which the the two nodes are moving away from each other by merely observing the slope of the progression. The threshold to trigger a prediction is a tunable parameter that would determine the *look-ahead time* for the link failure. The objective is to enable the completion of an alternate route computation before the failure of the current path. A critical aspect of the prediction process is the propagation model used. Since the two-ray ground reflection model is assumed for distances greater than 100m, the corresponding equation is used to calculate the threshold receive power corresponding to a particular slope (and hence speed) and look-ahead time. Based on the model, the received power P_r can be specified as:

$$P_r = \frac{K}{d^4}$$

where K is a constant and is a function of the receive and transmitter antenna gains and heights, and the transmit power. d is the distance between the transmitter and the receiver. Thus given a look-ahead time l and the observed relative speed between the two nodes s, the threshold power to trigger a prediction can be calculated as:

$$P_{RFP} = \frac{K}{r - s * l}$$

where r is the transmission range. The speed s is computed from the slope of the history of transmission powers observed. If the size of the history is N (N packet reception powers and corresponding times), the speed is calculated as follows:

$$s = \frac{\sum_{i=1}^{N} \frac{P_{i+1} - P_i}{t_{i+1} - t_i}}{N - 1}$$

where P_i and t_i represent the received power and time of reception for the ith packet in the history. When a source receives a predicted route failure message it issues a new route request, but continues to use the current path either till the new route is computed or the current route fails and it receives a normal route error. Route requests are suppressed based on the same thresholds used to predict route failures. Hence, a route that is close to failure will not be chosen during any route computation process.

In essence, the mechanism tries to predict link failures in advance and proactively determine an alternate route before the failure of the route currently being used. This would prevent the retransmission timers at the TCP sender from firing out during route computations and leading to a degradation in throughput or in the worst case resulting in connection stalls due to repeated timeouts. The prediction mechanism is merely a heuristic and can fail either by predicting link failure wrongly or failing to predict an actual link failure. In the first scenario, the throughput of the corresponding connection is left unaffected since the source will continue to use the current path until a new alternate path is computed or the current path fails. Since the current path will not fail, the source will switch its connection only upon the recomputation of an alternate path. Further, if the current path is the best path, it will again be recomputed as the alternate path thus preventing any sub-optimality because of wrong predictions. The drawback of wrong predictions is the route recomputation overhead that is incurred. On the other hand, if a route failure is not predicted successfully, the performance of the connection will be only as bad as the scenario wherein no prediction mechanism is employed.

Finally, the prediction mechanism will successfully predict only mobility related route failures. Other possibilities such as congestion based route failures will not be captured by the prediction mechanism and will trigger normal route errors as usual.

- *Proactive Route Errors (PRE):*

  If a link failure occurs either due to congestion, or due to mobility but has not been successfully predicted by the prediction mechanism, the third mechanism in *Atra* tries to minimize the latency involved in the route failure information being carried to the source(s) using the link of failure. In the default set-up, DSR will issue a route error to only the source of the packet that triggered the link failure detection at the MAC layer. If multiple sources are using the same link, packets will have to arrive from them at that node before route errors will be sent back to them increasing the latency between the link failure detection and the time at which the sources are informed. This latency is further inflated because subsequently arriving packets from other sources will have to go through the *MAC failure detection time* cycle before the link failure is inferred. However, in *Atra* each node maintains a cache of the source identifiers of TCP connections that have used a particular link in the past T seconds. When a link failure is detected, all sources that have used the link in the past T seconds are informed about the link failure through normal route errors. This reduces the latency involved in the route failure information delivery which consequently reduces the number of losses and also triggers earlier alternate route computations.

The proactive route error mechanism can prove to be disadvantageous when a link failure has occurred due to congestion. Consider an example where a link between nodes A and B is traversed by 2 TCP connections $f1$ and $f2$. In the default set-up, when a packet belonging to $f1$ experiences congestion related link failure, only $f1$ would be informed of the link failure prompting $f1$ to choose a different route and thus relieving congestion along the original path for $f2$. However, when the proactive route error mechanism is used, both $f1$ and $f2$ will be informed of the route failure making both of them to recompute their routes (although the same path might be chosen all over again). However, the characteristic of the default set-up to let route requests through in preference to data packets results in routes being chosen irrespective of the congestion along the path. Hence, in the example considered there is nothing to prevent flow $f1$ from choosing the same path again even under the default set-up.

Performance

While the symmetric route pinning mechanism reduces the probability of route failures, the route failure prediction mechanism reduces the occurrence of route failures by predicting them proactively. These two mechanisms directly reduce the number of retranmission timeouts in TCP. However, when a route failure does occur, the proactive route error mechanism reduces the delay in informing the sources of the route failure, thereby reducing the probability of a retransmission timeout. Hence, these three mechanisms in concert reduce the number of retransmission timeouts experienced by the connection as observed in Figure 5.7(a) when compared to the default TCP case in Figure 5.2(c) for the 1 connection scenario. A direct benefit of the reduction in the number of timeouts, is the resulting reduction in the loss percentage of packets and the consequent increase in throughput as observed in Figures 5.7 (b) and (c) respectively.

Trade-offs

The lower layer mechanisms in the *ATRA* framework help improve TCP's performance without requiring any changes to the transport layer, by appropriately taking actions to mask out the negative impacts of route failures caused due to mobility. However, this is achieved at the cost of lower layers being required to be TCP-aware in their operations. Also, the route prediction mechanism relies on the signal strength of the received packets, it is challenge to make such predictions accurate under conditions of signal fading due to obstacles, multipath, etc. Finally, several characteristics of TCP that are by themselves inappropriate for operation over ad-hoc networks, are not addressed by this framework.

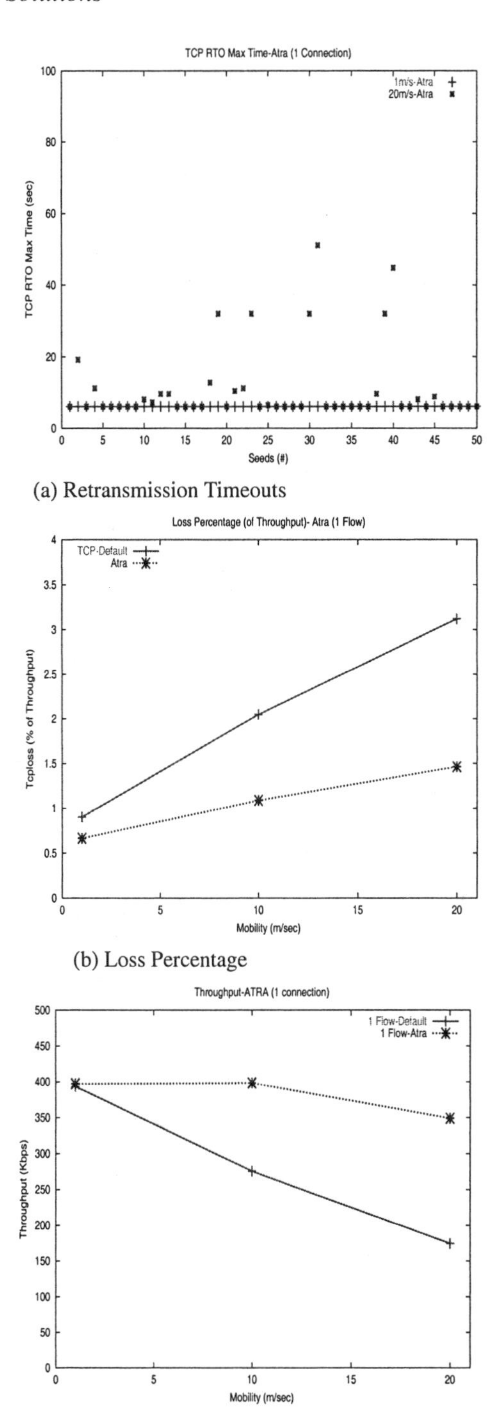

(a) Retransmission Timeouts

(b) Loss Percentage

(c) Throughput

Figure 5.7. Atra (1 Flow)

Related Work

There are other research works that have attempted to identify factors affecting TCP's performance in cellular and multi-hop wireless network scenarios [2, 3]. [2] studies the effect of routing and link layer mechanisms on TCP performance. It investigates cache management strategies and discusses the effect of link layer re-transmissions on TCP throughput. The study is conducted on a static wireless network. [3] investigates the impact of MAC protocol on the performance of TCP in multi-hop wireless networks.

5.6 Ad-hoc Transport Protocol

The last approach that we discuss in this chapter involves a complete redesign of the transport layer protocol. Though the degree of change in the transport layer behavior through such a design is significant, the advantage is that the designed protocol can be better tuned to the nuances and characteristics of the target environment. A transport protocol that has adopted this approach is the *Ad-hoc Transport Protocol* (ATP) proposed by Sundaresan et al.

The design of ATP is an anti-thesis of TCP's design. Its different components are designed in such a manner so as to solve the problems experienced by TCP over ad-hoc networks. ATP is a rate based transport protocol with functonalities that reside at one of three entities, namely ATP sender, intermediate nodes and ATP receiver. The ATP sender is responsible for connection management, reliability, congestion control, and initial rate estimation. The intermediate nodes assist the sender in its operations by providing network feedback with respect to congestion control and initial rate estimation. The ATP receiver is responsible for collating the feedback information provided by the intermediate nodes before sending the final feedback to the ATP sender for reliability, rate and flow control.

Mechanisms

A brief outline of the functionalities of ATP are:

- *Congestion Control:*

  Every intermediate node maintains two parameters, Q_t the average queuing delay experienced at the node, and T_t the average transmission delay at the node. Every packet generated from the source has a rate feedback field D in its header. Whenever a packet is to be transmitted at an intermediate node, the value D in the packet is compared with the sum of the parameters Q_T and T_t calculated with respect to the current packet. If the sum happens to be larger, the value D is replaced with this new value. When the packet reaches the receiver, the receiver performs an exponential averaging before sending the rate feedback D_{avg} to the sender in its feedback packet. When the sender obtains the feedback packet, it compares its current sending rate S with the feedback rate $\frac{1}{D_{avg}}$ (R) and accordingly performs congestion control. The congestion control operation has three phases: increase ($S < R - \phi S$), decrease ($S > R$) and maintain. Interested readers can find details on the congestion control operations in [13].

Since the intermediate nodes are involved in the congestion control operations, the feedback provided (delay based feedback) about the available bandwidth along the path is more accurate. This helps the rate adaptation mechanism function efficiently. Furthermore, the maintain phase is unique to ATP's congestion control mechanism where the sender maintains the rate without any fluctuations once the available bandwidth along the path has been probed. This is in contrast to TCP's act of probing for more bandwidth unless a loss is experienced and in turn helps make use of network resources more efficiently.

- *Initial Rate Estimation:*

  ATP uses a mechanism called *quick start* for initial rate estimation. In quick start *probe* packets are sent at a periodic interval to elicit feedback rate from the receiver. This mechanism probes for the available bandwidth along the path within a single *rtt* whereas TCP's slow start takes several *rtt's*. This helps the connection ramp up to the available network bandwidth within a short duration, thereby making better use of the network resources. Since the available bandwidth along the path is not known both during connection initiation and during underlying path changes, the quick start is performed in both these cases.

- *Reliability and Flow Control:*

  ATP receiver uses SACK blocks to provide loss information to the sender. The feedback is provided on a periodic basis to help keep the reverse path overhead small. The sender uses this SACK information by means of a scoreboard data structure as in TCP-SACK. However, since the sender does not use retransmission timer, the receiver has to always provide loss information starting from the first hole. ATP's Flow control and connection management are similar to the mechanisms in TCP.

The highlights of ATP that overcome the potential drawbacks of using TCP over ad-hoc networks can be summarized as follows: ATP uses rate based transmissions instead of window based transmissions and hence avoids the negative impacts arising due to burstiness. It uses *quick start* instead of slow start, which prevents the under- utilization of resources and also helps the fairness properties of the protocol. Delay is used as an indicator of congestion instead of loss. The is due to the inability of the MAC layer to distinguish the cause of the loss (congestion, random wireless errors or mobility-induced), which makes loss an inappropriate indicator of congestion. ATP uses a three phase rate adaptation mechanism instead of the LIMD mechanism of TCP. This is because connections in ad-hoc networks are vulnerable to route failures. Hence TCP's multiplicative decrease is unwarranted during route failures where most of the time a new route is chosen. Furthermore, the linear increase causes slow convergence to the optimal operating bandwidth. Finally, the coarse grained receiver feedback in ATP eliminates the data connection's dependence on ACKs and thereby the strong coupling between the forward and reverse paths.

Performance

The results presented in Figure 5.8 are for a single connection scenario. ATP's rate based transmissions eliminate the negative impacts resulting from burstiness of packet transmissions. This can be observed from the sequence number progression for the ATP flow in Figure 5.8(a) where the packet transmissions are more uniformly spaced out when compared to the bursty transmissions in the case of default TCP in Figure 5.2(a) for the same scenario. ATP overcomes the under-utilization of network resources, resulting from the use of slow- starts in ad-hoc networks, by employing the quick start mechanism. Though, the quick start mechanism probes for the available network bandwidth along the path within a single *rtt*, it is still a bandwidth estimation phase and hence we are interested in the under-utilization of network resources resulting from the use of quick start. The result in Figure 5.8(b) indicates that the total amount of time spent by the ATP connection in the quick start phase is an order of magnitude less than that spent by a default TCP flow in slow start for the same scenario. Finally, the various design elements of ATP help it obtain a significant throughput improvement over the default TCP case, as shown in Figure 5.8(c).

Trade-offs

ATP, being a protocol tailored to the characteristics of ad-hoc networks, attains significant performance improvement. However, its inter-operability with TCP is of prime concern if a mobile host using ATP also wants to be a part of the static Internet. The inter-operability of ATP with TCP is not evident, and is currently being investigated [13]. Further, while the strong coupling between the data and ACK paths in TCP is alleviated by the use of coarse-grained receiver feedback, ATP's rate adaptation mechanism still relies on receiver feedback, and the tuning of ATP's receiver feedback rate is still not fully addressed [13].

Related Work

An example of another protocol whose design is tailored to a specific target environment is the satellite transport protocol (STP) [14]. It has been designed

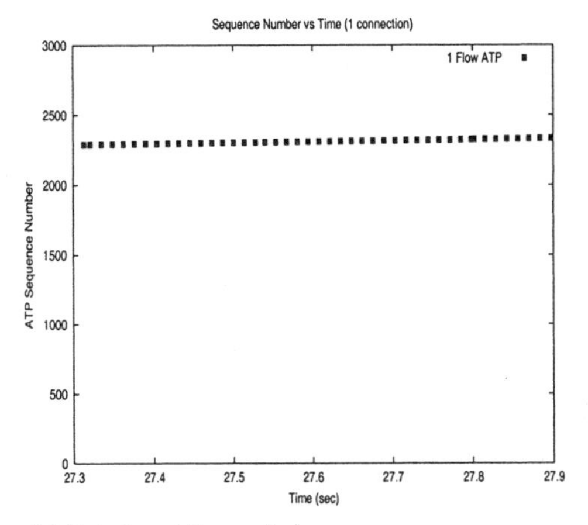

(a) Rate-based Transmissions

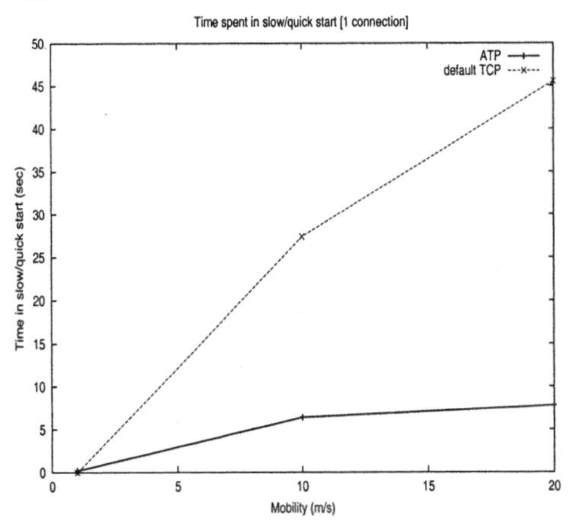

(b) Time Spent in Quick/Slow-start

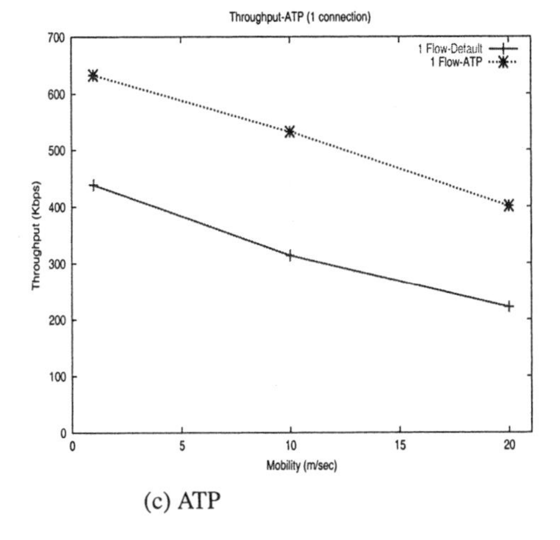

(c) ATP

Figure 5.8. ATP (1 Flow)

Modified TCP: TCP-ELFN	TCP-aware Cross-layered Solutions: ATRA	Ad-hoc Transport Protocol: ATP
Explicit link failure notification from network entities hides latency of route re-computations from TCP sender	Comprises of 3 mechanisms at the routing and MAC layers	Rate based transmissions replace window based transmissions, alleviating the negative impacts of burstiness
A "stand-by" mode for TCP sender to "freeze" state on receiving ELFN message from the routing layer	Symmetric Route Pinning at the routing layer to reduce the probability of route failures	Quick start, instead of slow start, is used to prevent under-utilization of resources
Periodic "probing" of the network during "stand-by" mode to determine establishment of a route	Route Failure Prediction at the MAC layer to help sources proactively obtain an alternate route on predicting a route failure	Delay serves as congestion indicator; a three phase rate adaptation mechanism re-places TCP's inappropriate LIMD mechanism
Restoration of network state on re-establishment of route, based on the state maintained before "freeze", helps utilize network resources efficiently	Proactive Route Error at the routing layer to reduce the latency involved by source in detecting a route failure by informing 'all' the sources using the failed link	Coarse grained receiver feedback reduces the dependence of data path on ACK path

Figure 5.9. Key Elements in Approaches

for operation in a datagram-based satellite data network. STP's automatic repeat request (ARQ) mechanism uses only selective acknowledgments. This helps STP use significantly less bandwidth on the return path. The STP sender retransmits only those specific packets that have been explicitly requested by the receiver and hence does not use any *timeouts*. STP is also relatively insensitive to variations in *rtt*, making it an attractive candidate for asymmetric networks. In a different context, the NACK-oriented reliable multicast protocol (NORM) [15] provides NACK based reliability and is similar to the ATP protocol, since it employs rate based transmissions.

5.7 Summary

The focus of this chapter is to investigate the problems experienced by the TCP transport layer protocol, when operating over ad-hoc wireless networks. To this end, the key design components of TCP are first highlighted, and for each of the components, the relevance, or lack there-of, with respect to the characteristics of ad-hoc networks was discussed. Then, three major categories of approaches to improve transport layer performance over ad-hoc networks are identified, and an instance of each category is discussed in detail. The trade-offs that stem from the adoption of any one solution is also discussed.

References

[1] Co-operative Association for Internet Data Analysis (CAIDA), "http://www.caida.org/," .

[2] G. Holland and N. H. Vaidya, "Impact of Routing and Link layers on TCP Performance in Mobile Ad-hoc Networks," in *Proceedings of IEEE WCNC*, New Orleans, September 1999.

[3] M. Gerla, K. Tang, and R. Bagrodia, "TCP Performance in Wireless Multi Hop Networks," in *Proceedings of IEEE WMSCA*, New Orleans, Feb 1999.

[4] M. Patel, N. Tanna, P. Patel, and R. Banerjee, "TCP over Wireless Networks: Issues, Challenges and Survey of Solutions," .

[5] C. E. Koksal and H. Balakrishnan, "An Analysis of Short-term Fairness in Wireless Media Access Protocols (poster)," in *Proceedings of ACM SIGMETRICS, Measurement and Modeling of Computer Systems*, Santa Clara, CA, 2000, pp. 118–119.

[6] D. Johnson, D.A. Maltz, and J. Broch, "The Dynamic Source Routing Protocol for Mobile Ad Hoc Networks ," in *MANET Working Group. IETF, Internet Draft, draft-ietf-manet-dsr- 07.txt*, Feb 2002.

[7] C. E. Perkins and E. M. Royer, "Ad-hoc On-demand Distance Vector (AODV) Routing," in *MANET Working Group. IETF, Internet Draft, draft-ietf-manet-aodv-12.txt*, Nov 2002.

[8] J. P. Monks, P. Sinha, and V. Bharghavan, "Limitations of TCP-ELFN for Ad hoc Networks," in *Workshop on Mobile and Multimedia Communication*, Marina del Rey, CA, Oct. 2000.

[9] K. Chandran, S. Raghunathan, S. Venkatesan, and R. Prakash, "A Feedback Based Scheme for Improving TCP Performance in Ad-Hoc Wireless Networks," in *Proceedings of International Conference on Distributed Computing Systems*, Amsterdam, May 1998, pp. 472–479.

[10] T. D. Dyer and R. Bopanna, "A Comparison of TCP Performance over Three Routing Protocols for Mobile Ad Hoc Networks ," in *Proceedings of ACM MOBIHOC 2001*, Long Beach, CA, Oct 2001.

[11] J. Liu and S. Singh, "ATCP: TCP for Mobile Ad Hoc Networks," in *IEEE Journal on Selected Areas in Communications*, 2001.

[12] V. Anantharaman and R. Sivakumar, "A Microscopic Analysis of TCP Performance Analysis over Wireless Ad Hoc Networks," in *Proceedings of ACM SIGMETRICS 2002. (Poster Paper)*, Marina del Rey, CA, June 2002.

[13] K. Sundaresan, V. Anantharaman, H-Y. Hsieh, and R. Sivakumar, "ATP: A Reliable Transport Protocol for Ad-hoc Networks ," in *Proceedings of ACM MOBIHOC 2003*, Annapolis, MD, Jun 2003.

[14] T. Henderson and R. Katz, "Satellite Transport Protocol (STP): An SSCOP-based Transport Protocol for Datagram Satellite Networks," in *Proceedings of 2nd Workshop on Satellite-Based Information Systems (WOSBIS)*, Budapest, Hungary, 1997.

[15] M. Handley, C. Bormann, B. Adamson, and J. Macker, "NACK Oriented Reliable Multicast (NORM) Protocol Building Blocks," in *Internet Draft, RMT Working Group, draft-ietf-rmt-bb-norm-05.txt*, March 2003.

Chapter 6

ENERGY CONSERVATION

Robin Kravets
Department of Computer Science
University of Illinois at Urbana-Champaign
rhk@uiuc.edu

Cigdem Sengul
Department of Computer Science
University of Illinois at Urbana-Champaign
sengul@uiuc.edu

Abstract Energy is a limiting factor in the successful deployment of ad hoc networks since nodes are expected to have little potential for recharging their batteries. In this chapter, we investigate the energy costs of wireless communication and discuss the mechanisms used to reduce these costs for communication in ad hoc networks. We then focus on specific protocols that aim to reduce energy consumption during both active communication and idle periods in communication.

Keywords: Communication-time energy, idle-time energy, power control, topology control, energy-aware routing, suspend/resume scheduling, power management.

Introduction

The limited energy capacity of mobile computing devices has brought energy conservation to the forefront of concerns for enabling mobile communications. This is a particular concern for mobile ad hoc networks where devices are expected to be deployed for long periods of time with limited potential for recharging batteries. Such expectations demand the conservation of energy in all components of the mobile device to support improvements in device lifetime [11] [10] [25] [38] [42] [35]. In wireless networks, there is a direct tradeoff between the amount of data an application sends and the amount of energy consumed by sending that data. Application-level techniques can be used to reduce the amount of data to send, and so the amount of energy consumed. However, once the application decides to send some data, it is up to the network to try to deliver it in an energy-efficient manner. To support energy-efficient communication in ad hoc networks, it is necessary to consider energy consumption at multiple layers in the network protocol stack. At the network layer, intelligent routing protocols can minimize overhead and ensure the use of minimum energy routes [7] [19] [41] [58] [60] [61]. At the medium access control (MAC) layer, techniques can be used to reduce the energy consumed during data transmission and reception [14] [30] [45] [31] [44] [70]. Additionally, an intelligent MAC protocol can turn off the wireless communication device when the node is idle [26] [34] [56] [57] [65] [69] [72] [35].

Communication in ad hoc networks necessarily drains the batteries of the participating nodes, and eventually results in the failure of nodes due to lack of energy. Since the goal of an ad hoc network is to support some desired communication, energy conservation techniques must consider the impact of specific node failures on effective communication in the network. At a high level, achieving the desired communication can be associated with a definition of network lifetime. Current definitions of network lifetime include: 1) the time when the first node failure occurs [5], 2) the fraction of nodes with non-zero energy as a function of time [22] [67] [68], 3) the time it takes the aggregate delivery rate to drop below a threshold [8], or 4) the time to a partition in the network. In the context of any of these definitions, it may also be useful to consider node priority in the definition of lifetime. For example, the network lifetime could be defined as the time the first high priority node fails. In general, one static definition of lifetime does not fit all networks. In this chapter, we do not discuss the impact of the definition of network lifetime or node failures due to depleted batteries on the communication in the network. Instead, we present approaches to energy conservation that minimize energy consumption for communication in ad hoc networks. However, these approaches can be tuned to support the desired communication and the definition of network lifetime as needed by the specific ad hoc network.

Energy conservation can be achieved in one of two ways: saving energy during active communication and saving energy during idle times in the communication. The first targets the techniques used to support communication in an ad hoc network and is typically achieved through the use of energy-efficient MAC and routing protocols. The second focuses on reducing the energy consumed when the node is idle and not participating in communication by placing the node in a low-power state. In this chapter, we first define the costs associated with communication in ad hoc networks and then discuss the use of communication-time and idle-time energy conservation.

Card	Transmit	Receive	Sleep
Cisco Aironet 350	2.25W	1.35W	75mW
Linksys wpc11	1.42W	462mW	297mW
ORiNOCO 11b	1.43W	925mW	45mW
Socket Low Power SDIO	924mW	49.5mW	49.5mW
Mica2 Mote	89.1mW	33mW	$< 1\mu A$

Table 6.1. Transmit, receive and sleep mode energy costs for selected wireless cards.

6.1 Energy Consumption in Ad Hoc Networks

In general there are three components to energy consumption in ad hoc networks. First, energy is consumed during the transmission of individual packets. Second, energy is consumed while forwarding those packets through the network. And finally, energy is consumed by nodes that are idle and not transmitting or forwarding packets. To understand how and when energy is consumed in ad hoc networks, it is necessary to consider these costs for data packets forwarded through the network and for control packets used to maintain the network. To lay the groundwork for discussing energy efficient communication protocols in ad hoc networks, we define these costs for communication and introduce energy-saving mechanisms used by many protocols.

6.1.1 Point-to-Point Communication

The basis for all communication in ad hoc networks is the point-to-point communication between two nodes. At each node, communication impacts energy consumption in two ways. First, the wireless communication device consumes some base energy when it is activated and idle (see Table 6.2. Note that specifications for most current wireless devices do not provide a differentiation between idle and receive costs). Second, the act of transmitting a packet from one node to another consumes energy at both nodes. Transmission energy is determined by the base transmission costs in the wireless card (see Table 6.1) and the transmit power level at the sender (see Table 6.2). Reception energy depends on the base reception costs in the wireless card and the processing costs for reception (see Table 6.1). The amount of time needed for the packet transfer determines the amount of time the card must be active, and so directly determines the energy consumed by the base card costs for both transmission and reception. This time is determined by two factors: the control overhead from packet transmission and the rate at which the packet is transmitted.

The per-packet control overhead is determined by the mechanisms of the medium access control (MAC) protocol. Depending on the chosen protocol, some energy may be consumed due to channel access or contention resolution. For example, in IEEE 802.11 [26], the sender transmits an RTS (ready to send) message to inform the receiver of the sender's intentions. The receiver replies

Card	Transmit Power Levels
Cisco Aironet 350	100, 50, 30, 20, 5 mW
Socket Low Power SDIO	max 25 mW

Table 6.2. Transmit power levels for selected wireless cards with power control capabilities.

Card	Rates
IEEE 802.11 b,g	11, 5.5, 2, 1 Mbps
IEEE 802.11 a,g	54, 48, 36, 24, 18, 12, 9, 6 Mbps
Mica2 Motes	12 Kbps

Table 6.3. Transmission rates for selected wireless card types.

with a CTS (clear to send) message to inform the sender that the channel is available at the receiver. The energy consumed for contention resolution includes the transmission and reception of the two messages. Additionally, the nodes may spend some time waiting until the RTS can be sent and so consume energy listening to the channel. In this chapter, we focus on the use of RTS/CTS-based protocols. While it has been shown that such protocols may not be optimal for throughput [37], there is no widely accepted alternative for communication in mobile ad hoc networks.

Once channel access and contention resolution have determined that a packet may be sent, many wireless network cards provide multiple rates at which the data can be transmitted, which determines the time needed to send the data (See Table 6.3). The specific transmission rate used is determined by a number of factors, including the signal-to-noise ratio (SNR) and the target reliability of the transmission [19] [41] [58] [60]. In general, the signal strength at the receiver, which determines the SNR, varies directly with the sender's transmit power level and varies inversely with the distance between the sender and the receiver. This relationship can be formulated as:

$$ReceiveSignal \propto TransmitPower/Distance^n, \tag{1.1}$$

where the path loss exponent n varies from 2 to 6 [51], although is most commonly used as 2 or 4. For the receiver to correctly receive the packet, the SNR must be over a certain threshold. As long as the receive SNR is maintained above this threshold, the transmit power level at the sender can be reduced, directly reducing energy consumption at the sender. The adaptation of the sender's transmit power level is called *power control* and is the main tool used to conserve energy during active communication. For the remainder of this chapter, we use power level to mean transmit power level.

Finally, energy is consumed to compensate for lost packets, generally via some number of retransmissions of the lost packets. While reliability is generally the domain of the transport layer, the MAC layer in most wireless devices compensates for some packet failure by retransmitting the packet up to some retransmit limit number of times before considering the packet lost. For current energy conserving protocols, this cost is only considered by protocols that aim to avoid low quality channels and so avoid needing to retransmit packets.

6.1.2 End-to-End Communication

End-to-end communication in ad hoc networks is supported by all nodes participating in route maintenance and data forwarding. Therefore, network-wide energy consumption includes any control overhead from routing protocols, including route setup, maintenance and recovery, as well as the impact of the chosen routes on the energy consumed at the intermediate nodes to forward data to the receiver. The choice of a specific route is determined by the metrics used in the routing protocol. Initial protocols use hop count as a primary metric [29] [47], although delay often implicitly impacts route choices [29]. More recent protocols suggest the use of extended metrics such as signal strength [12], stability [63] and load [36] [46], all of which impact performance and so implicitly impact energy consumption [18]. Energy can also be used explicitly to choose routes that minimize energy consumption [54] [64] or avoid nodes with limited energy resources [58] [33]. Additionally, when a route breaks, it is essential to use energy-efficient mechanisms to find a new route, avoiding a reflooding of the network whenever possible. At the network layer, *energy-efficient routing protocols* combine these techniques with power control for additional energy conservation during active communication.

6.1.3 Idle Devices

A wireless communication device consumes energy when it is idle or listening to the channel (See receive costs in Table 6.1). Such idle costs can dominate the energy consumption of a node, especially if there is not much active communication. Idle-time energy conservation can be achieved by suspending the communication device (i.e., placing it in a low-power mode). Low-level management of *device suspension* is generally handled in the MAC layer. Such power-save modes monitor local communication to determine when a device can be suspended (i.e., no immediate communication) and when it should be awake to communicate with its neighbors. While energy is conserved in these power-save modes, there is a limitation placed on the communication capacity of the network since all communication to and from the node is suspended. Higher layer *power management protocols* trade off energy and performance by determining when to transition between power-save mode and standard active mode.

6.1.4 Energy Conservation Approaches

Once all of these costs are understood, two mechanisms affect energy consumption: power control and power management. If these mechanisms are not used wisely, the overall effect could be an increase in energy consumption or reduced communication in the network. The remainder of this chapter is broken into two sections. We first present techniques for communication-time energy conservation, focusing on the impact of power control and energy-efficient routing. We follow this with a presentation of idle-time energy conservation techniques, looking at both low level suspend/resume mechanisms and higher level power management.

6.2 Communication-Time Energy Conservation

The goal of communication-time energy conservation is to reduce the amount of energy used by individual nodes as well as by the aggregation of all nodes to transmit data through the ad hoc network. Two components determine the cost of communication in the network. First, direct node-to-node transmissions consume energy based on the power level of the node, the amount of data sent and the rate at which it is sent. The amount of data is determined by the application and the rate is determined by the characteristics of the communication channel. Although the transmission rate can also be adapted by the sender [23], we do not consider such rate control in this chapter. However, the power level can be controlled by the node to reduce energy consumption. Such *power control* must be performed in a careful manner since it can directly affect the quality and quantity of communication in the network. Second, energy is consumed at every node that forwards data through the network. Such costs can be minimized using *energy-aware routing* protocols. This section first discusses the use of power control and its impact on communication in ad hoc networks. We then present power control protocols and energy-aware routing protocols that aim to minimize energy consumption for communication in the network.

6.2.1 Power Control

Current technology supports power control by enabling the adaptation of power levels at individual nodes in an ad hoc network. The power level directly affects the cost of communication since the power required to transmit between two nodes increases with the distance between the sender and the receiver. Additionally, the power level defines the communication range of the node (i.e., the neighbors with which a node can communicate), and so defines the topology of the network. For devices capable of power control, the power level can be adapted up to a *transmit power level threshold*, as defined by the capabilities of the device (see Table 6.2). This threshold defines the maximum energy cost for communication. Due to the impact on network topology, artificially limiting the power level to a *maximum transmit power level* at individual nodes is called *topology control*. Topology control protocols adapt this maximum within the constraints of the threshold to achieve energy-efficient communication by limiting the maximum cost of a transmission. The impact of power control on communication is twofold. First, adjusting power levels affects channel reservation. Second, power control determines the cost of data transmission.

During channel reservation, the power level directly defines the physical range of communication for a node and the physical area within which channel access control must be performed. Given the shared characteristics of wireless communication channels, any node within transmission range of the receiver can interfere with reception. Similarly, the sender can interfere with reception at any node within its transmission range. Therefore, MAC layer protocols coordinate all nodes within transmission range of both the sender and the receiver. In the context of RTS/CTS-based protocols, the channel is reserved through the transmission of RTS and CTS messages. Any other node that hears these messages backs off, allowing the reserving nodes to communicate undisturbed. The power level at which these control messages are sent defines the area in which other nodes are silenced, and so defines the spatial reuse in the network [20] [24] [37] [62]. Since topology control determines the maximum power level for each node in the network, topology control protocols that minimize power levels increase spatial reuse, reducing contention in the network and reducing energy consumption due to interference and contention.

The use of power control can result in nodes with different maximum power levels. While utilization of heterogeneous power levels increases the potential capacity of the network, it increases the complexity and degrades the effectiveness of the control protocols. Therefore, it is necessary to understand these trade-offs to decide whether to allow heterogeneous power levels or to require all nodes to use the same maximum power level.

In a random uniformly distributed ad hoc network where traffic patterns are optimally assigned and each transmission range is optimally chosen, the maximum achievable throughput is $O(\frac{1}{\sqrt{n}})$ for each node, where n is the number of nodes in the network [21]. When a homogeneous, or common, power level is used (i.e., without optimal heterogeneous power level assignments), the achievable throughput closely approaches this optimum [32]. Therefore, common power can be effective in such networks. However, the results for common power in uniformly distributed networks are not applicable to non-uniformly distributed networks [20]. To maintain connectivity in a network where nodes are clustered, the common power approach converges to higher power levels than the heterogeneous approach, sacrificing spatial reuse and energy.

While heterogeneous power levels can improve spatial reuse, the mechanisms used for channel reservation are compromised, resulting in asymmetric links

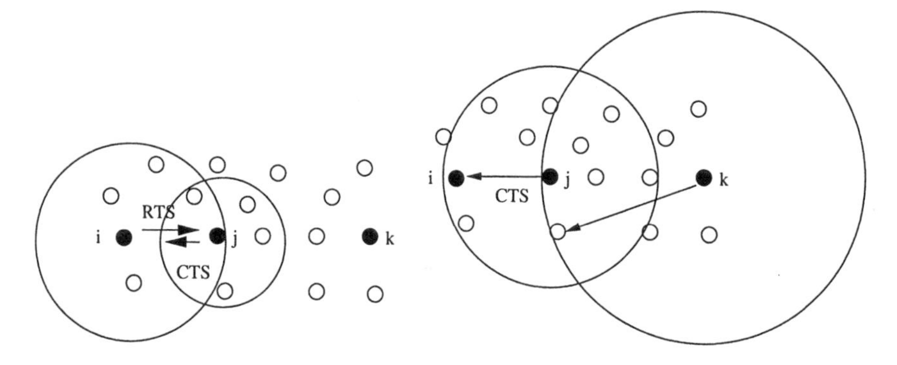

Figure 6.1. Node j's power level is less than node i's and communication is not possible.

Figure 6.2. Node j's CTS does not silence node k, and so node k can interfere with node j, since node k's power level is higher.

(see Figure 6.1) and in more collisions in the network [30]. For a homogeneous network where all nodes transmit with identical power levels, RTS/CTS-based protocols, such as IEEE 802.11, achieve contention resolution while limiting the occurrence of collisions. However, in a heterogeneous network where each node is capable of transmitting with different power levels, collisions may occur if a low-power node attempts to reserve the channel with an RTS message that is not heard by high-power neighbors that are close enough to disrupt communication [48] (See Figure 6.2). Therefore, control message transmission should use the threshold power level, leaving little potential for additional spatial reuse. PCMA [43] suggests the use of a second channel to transmit a busy tone, allowing senders to monitor the strength of the busy tone signal to dynamically determine a maximum power level that would not interfere with ongoing communication. However, PCMA was designed in the context of single hop wireless networks and it is yet unclear how to apply it to multihop wireless networks. Although channel reservation for nodes with heterogeneous power levels has not yet been solved in the context of ad hoc networks, future protocols may enable better channel reservation. Therefore, we discuss topology control protocols for both homogeneous and heterogeneous networks.

Once the communication range of a node has been defined by the specific topology control protocol, the power level for data communication can be determined on a per-link or even per-packet basis. If the receiver is inside the communication range defined by the specific topology control protocol, energy can be saved by transmitting data at a lower power level determined by the distance between the sender and the receiver and the characteristics of the wireless communication channel [19] [41] [58] [60]. When limited to the transmission of data messages, we call such transmit power control *transmission control*. In the context of RTS/CTS-based protocols, transmission control can easily be used to limit power level adaptation to the transmission of data, leaving control message transmission at the maximum power level [19].

Although reducing the power level only during data transmission directly reduces the transmission energy consumption, it can cause more collisions in the network [30] [48]. If the same power level is used for both control and data messages, nodes that miss the control message exchange still back-off during the data transmission since they sense a busy channel. If the the data is sent at a lower power level, nodes that miss the control message exchange may not sense a busy channel and so could unintentionally interfere with the data transmission. To compensate for these collisions, PCM [30] uses the threshold power level to send the RTS and CTS messages and uses the minimum power level necessary to transmit the ACK. However, to send the DATA, PCM alternates between short transmissions at the threshold and longer transmissions at the minimum power level. These "pulses" at the threshold power level indicate to other nodes that there is active communication and the channel is already reserved. While saving energy by sending most of the data message at a lower power level, PCM does not enable any extra spatial reuse.

Senders can use transmission control with very little overhead. Transmission control can be supported in a fully localized manner since it only needs information about the state of the communication channel between the sender and the receiver. For example, in the context of an RTS/CTS-based protocol, the receiver can return the observed signal strength of the RTS in the CTS packet [27] [1]. The sender can use the received signal strength along with the original power level for the RTS to determine an optimal data power level [19] [41] [58] [60]. Energy-aware routing protocols can then use these optimized data transmission costs to find minimum cost routes through the network.

6.2.2 Topology Control

Topology control aims to reduce the maximum power level at individual nodes to minimize energy consumption and maximize spatial reuse while maintaining connectivity in the network. However, aggressive topology control can create a network that is easily partitioned by the loss or failure of one node. Fault tolerance can be improved by requiring the topology control protocol to find a graph where multiple node failures are required to cause a partition. Additionally, the majority of topology control protocols are designed for static networks, limiting their ability to maintain the network topology in the presence of mobility.

Topology control protocols can be divided into two types: *common power* and *heterogeneous power*. Common power protocols find the common maximum power level for all nodes and heterogeneous protocols choose a maximum power level for each node. We first present both common and heterogeneous topology control protocols and then discuss the impact of mobility on all protocols.

Common Power. When all nodes share the same maximum power level, this common power should be chosen as small as possible to limit the maximum energy consumption and to achieve high spatial reuse. A common power that is too high increases the number of neighbors at a given node, which increases the number of nodes that can cause interference at that node, increasing energy consumption and reducing spatial reuse. On the other hand, if the common power is too low, the network may be disconnected, limiting effective communication in the network.

Given a discrete set of power levels $[P_{min}, P_{thresh}]$, if the network is connected when all nodes use the threshold power level (i.e., P_{thresh}), COMPOW [45] finds the smallest common power P that ensures the network remains connected. For each power level P, $R(P)$ is the set of nodes that are connected to a distinguished node when all nodes use common power. Thus, $R(P)$ is the reachable set for a common power P. Since the network is connected at P_{thresh}, $R(P_{thresh})$ is the maximal reachable set. COMPOW finds the minimum power level P_{common} that maintains this maximal reachable set (i.e., $R(P_{thresh}) = R(P_{common})$). To find each $R(P)$, COMPOW runs one proactive routing protocol at each power level up to P_{thresh} to populate the neighbor sets for each node at each power level. The result is a minimum common power that achieves connectivity. However, there is no fault tolerance built into COMPOW and the failure of a critical node can partition the network.

If the network is not connected at P_{thresh}, COMPOW finds the minimum power level that maintains connectivity for every connected component of the network. In a network where nodes are clustered, the common power must be chosen to connect the clusters to each other and therefore may converge to a higher power level (see Figure 6.3). The CLUSTERPOW power control protocol [31] addresses this problem by choosing per packet power levels so that intra-cluster communication uses lower common power and only inter-cluster communication uses higher power levels (see Figure 6.3). This use of multiple power levels at the same time to reach different clusters is a step towards heterogeneous power control approaches, which are discussed next.

Heterogeneous Power. Allowing each node to pick its own maximum power level increases spatial reuse in the network and so increases network capacity. Heterogeneous power topology control protocols use local information to determine which links must be part of the network to maintain connectivity and set the power levels to ensure the presence of those links. We discuss four approaches to heterogeneous power topology control: Connected MinMax [50], Enclosure [52], Cone-Based [66] and Local Minimum Spanning Tree [40].

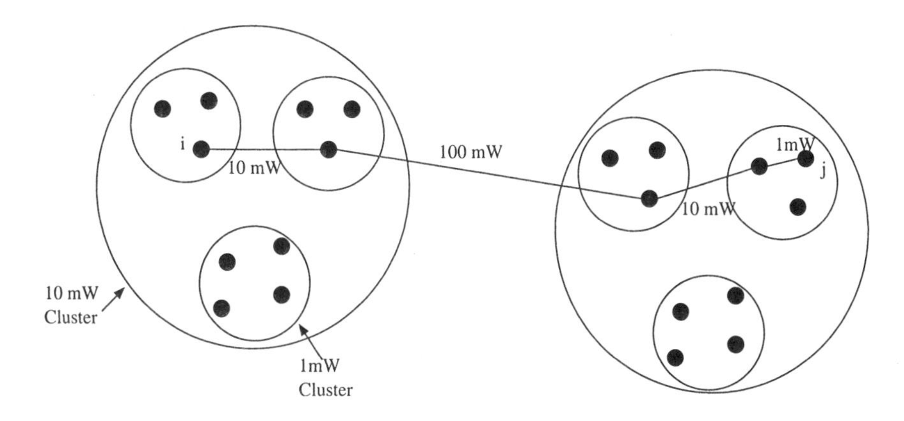

Figure 6.3. COMPOW computes a common power level of 100mW for the network, which shows that a common power level is not appropriate for non-homogeneous networks. With CLUSTERPOW, the network has three clusters corresponding to 1mW, 10mW and 100mW. The 100 mW cluster is the whole network. A 10mW-100mW-10mW-1mW route is used for node i to reach node j.

Connected MinMax Power. In the first approach, the problem of adjusting the power level of individual nodes to create a desired topology is formulated as a constrained optimization problem with connectivity and bi-connectivity as constraints and maximum power level as the optimization objective [50]. The goal of the MinMax Power algorithm is to find the minimum energy needed to maintain a connected (or bi-connected) topology by minimizing the power level of the node with the maximum power level.

The multihop wireless network is represented as $M = (N, L)$, where N is the set of nodes and L is the set of coordinates of node locations. This algorithm requires knowledge of node locations for correct operation. A *least-power* function $\lambda(d)$ defines the minimum power level required to transmit to a distance d, based on current channel conditions. $\lambda(d)$ is defined as:

$$\lambda(d) = \gamma(d(l_i, l_j)) + S, \tag{2.1}$$

where γ is a monotonically increasing propagation function of the geographical distance between the location of node i, l_i, and the location of node j, l_j, and S is the receiver threshold, which determines the threshold signal strength needed for reception. S is assumed to be a known fixed cost for all nodes and, therefore, $\lambda(d)$ does not include the effects of channel fading and shadowing.

The MinMax Power algorithm finds a minimum energy topology that maintains connectivity in the network. For this optimization, a network forms a graph $G = (V, E)$, where V is the set of vertices corresponding to nodes and E is the set of edges corresponding to bi-directional links between nodes based on the maximum power level of the nodes. To improve fault-tolerance, the

MinMax Power algorithm can support more than minimum connectivity. A graph is k-vertex connected if and only if there are k vertex-disjoint routes between every pair of vertices. Therefore, the minimum power level assignment problem to achieve a connected ($k = 1$) and bi-connected ($k = 2$) multihop wireless network is formulated as follows [50]:

- *Connected MinMax Power:*

  Given a multihop wireless network $M = (N, L)$ and a least-power function $\lambda(d)$, find a per-node minimal assignment of power levels p such that M is 1-connected and $MAX_{i \in N}(p(i))$ is a minimum (i.e., the maximum power level assigned to any node $i \in N$ is minimized).

- *Bi-connectivity Augmentation with Minimum Power:*

  Given a multihop wireless network $M = (N, L)$, a least-power function $\lambda(d)$ and an initial assignment of per-node power levels p such that M is connected, find the per-node power level increase $\delta(i)$ such that the resulting graph is bi-connected (i.e., given a connected network, find the $p(i) + \delta(i)$ for each node $i \in N$ that makes the network bi-connected).

Given a static network and the location and least power function for all nodes, the above problems can be solved using the following polynomial (greedy) algorithms [50]. To find the power levels that connect the network, the CONNECT algorithm iteratively merges connected components until the whole network is connected. Initially, each node is an individual component. Node pairs are selected in non-decreasing order of their mutual distance. If the nodes are in different components, the power level of each node is increased to reach the other. This is continued until the whole network is one single component. Given a connected network and the power level assignments from the CONNECT algorithm, redundant links can be removed to ensure per-node minimums. The augmentation of a connected network to a bi-connected network is done via the BICONN-AUGMENT algorithm, which determines the bi-connected components in the network via a depth-first search. Node pairs are selected in non-decreasing order of their mutual distance and only joined if they are in different bi-connected components. This is continued until the whole network is bi-connected.

The Connected MinMax Power algorithm achieves the goal of a connected (or bi-connected) network that minimizes energy consumption. However, the algorithm has several limitations. First, both the CONNECT and BICONN-AUGMENT algorithms are centralized and require global information to construct the topology. Second, the construction requires location information, which can be expensive to collect and disseminate. Finally, the propagation model is quite simple and does not reflect the real characteristic of wireless communication such as shadowing or fading.

Enclosure Algorithm. The second approach uses a local optimization algorithm to find per-node maximum power levels that achieve minimum energy consumption [52]. This approach was designed for networks with a specific sink or master node that all other nodes want to communicate with. In this context, the enclosure algorithm focuses on multiple source - single destination communication.

To determine the maximum power level, each node creates a bounded region, called an enclosure, which defines the node's immediate neighborhood. All nodes inside the enclosure are direct neighbors and all nodes outside the enclosure are reached indirectly through neighbors. The enclosure is determined by finding relay regions associated with each neighbor, where indirect communication through neighbors with nodes in those regions is more power-efficient than direct communication. To calculate the enclosures, every node first broadcasts its location information at the threshold power level. A transmitting node i collects these broadcasts to determine the relay region R for each potential relay node j as follows:

$$R_{i \to j} \equiv \{(x, y) | P_{i \to j \to (x,y)} < P_{i \to (x,y)}\}, \tag{2.2}$$

where $P_{i \to j \to (x,y)}$ is the power level required to transmit from node i to a node at location (x, y) through the relay node j, and $P_{i \to (x,y)}$ is the power level required to transmit directly from node i to the node at location (x, y). The power consumption $P_{i \to j \to (x,y)}$ includes both transmit and receive power costs:

$$P_{i \to j \to (x,y)} = td_{i,j}^{n} + td_{j,(x,y)}^{n} + c, \tag{2.3}$$

where t is the minimum receive threshold at the receiver, d is the distance between i and j, path loss exponent $n \geq 2$ and c is the receive cost at the relay node j.

After determining all relay regions, node i can compose its enclosure and select its direct neighbors as those nodes that are not in any of the relay regions it has calculated. Node i then chooses a maximum power level that maintains connectivity to all of its neighbors. Figure 6.4 illustrates node i and five nodes $\{j, k, l, m, n\}$ it has discovered through the broadcast messages. Node i computes the relay regions for each of these nodes. The three regions computed for nodes j, k and l are illustrated in the figure. The bounded region around i is the enclosure of i. Node m falls in the relay region of node l (Relay Region 3 in the figure) and node n falls in the relay region of j (Relay Region 1 in the figure). Therefore nodes m and n do not belong to the enclosure of node i, which only maintains links to nodes j, k and l as its neighbors.

The *enclosure graph* of the network includes links that belong to all enclosures of all nodes in the network. The minimum power topology, which is a spanning tree with the master site as its root, is a sub-graph of the enclosure graph. A distributed Bellman-Ford shortest path algorithm [9] with

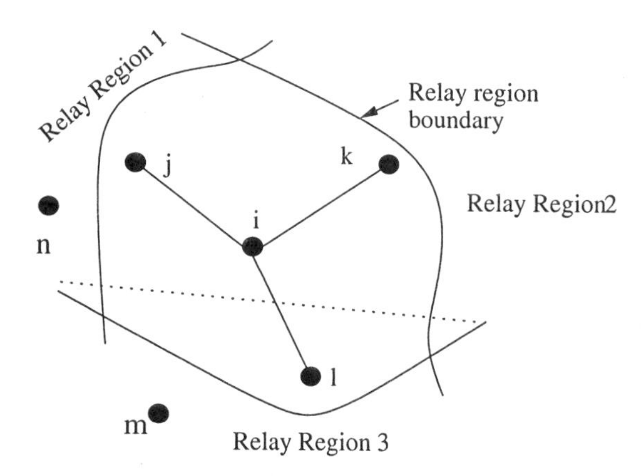

Figure 6.4. Enclosure of node i. Node i computes the relay regions of nodes j, k, and l. Relay Regions 1, 2 and 3 (corresponding to nodes j, k, and l respectively) specify the enclosure of node i. Node i maintains links only to nodes j, k and l. Nodes m and n are not contained in node i's enclosure, and therefore, are not its neighbors.

power consumption as the cost metric is used to find the minimum power paths from each node to the master site. The minimum-power topology is computed by simply removing all links from the enclosure graph that are not part of an energy-efficient shortest path.

The enclosure algorithm builds a strongly connected graph using only local information. It is guaranteed that there exists a path from any node i to any other node j, since the location of node j, (x_j, y_j) falls into the relay region of some neighbor of node i. However, the minimum power topology is computed for one destination and so cannot provide minimum energy communication for arbitrary communication between any two nodes. This approach could be extended to support such arbitrary communication by constructing minimum power topologies for all destinations. Additionally, nodes must be able to acquire their location information, since the algorithm must be able to determine the distances between nodes. Furthermore, the enclosure algorithm uses a fixed channel propagation model based on these distances to compute the relay regions. Such simple channel models do not capture the effect of noise levels at receivers, which may affect nodes differently. The use of this channel model will either be overly optimistic, causing some links to break, or overly pessimistic, causing some nodes to use a higher power level than necessary and wasting energy.

Cone-Based Topology Control. The third approach, cone-based topology control (CBTC), divides the space around each node into "cones" and attempts to create a link to at least one neighbor in every cone [66]. First, each node performs *neighbor discovery* by broadcasting discovery messages with increasing power levels until it has reached at least one neighbor in every cone of α degrees. This approach is limited to environments where the node can determine the direction of the sender when receiving a message. If no neighbor is reached in a particular cone even when transmitting at the threshold power level, that cone is left empty. A node's maximum power level is chosen as the minimum that maintains at least one neighbor in each cone, excluding the cones that had no neighbors at any power level. Figure 7.2 illustrates neighbor discovery by the cone-based algorithm for $\alpha = \pi/2$. In the figure, node i sets its power level to P_{max}, which maintains neighbors in cones I, II and IV. Since node j is out of receive threshold range, cone III is empty.

Once the initial power levels have been determined, nodes perform *redundant edge removal*, removing the edges that use more power than an indirect route. Specifically, node i removes an edge to node j if there exists a node k and:

$$p(i,j) + p(j,k) < p(i,k), \tag{2.4}$$

where $p(i,j)$ denotes the power required to send from node i to node j. From a performance point of view, a node should have as few neighbors as possible to reduce the contention and interference in its neighborhood. Therefore, it is desirable to remove some edges even if a direct transmission consumes less power than an indirect transmission. Therefore, Equation 2.4 is extended to:

$$p(i,j) + p(j,k) \leq q \times p(i,k), \tag{2.5}$$

where $q \geq 1$ is a constant that determines the threshold for edge removal even if a direct transmission is more power-efficient (for $q > 1$).

The resulting network constructed by the cone-based algorithm is connected for $\alpha \leq 5\pi/6$, if it is connected when all nodes transmit at the threshold power level [39]. Additionally, if $\alpha \leq 2\pi/3$, asymmetric edges can be removed while still maintaining network connectivity. This is not true for the case when $\alpha > 2\pi/3$, which requires adding a reverse edge for each asymmetric edge to preserve connectivity. We refer the readers to [39] for detailed proofs.

The cone-based algorithm depends only on directional information and does not assume that nodes have location information. However, current techniques for estimating direction without using location information require nodes to be equipped with multiple directional antennas, which can be more complex and consume more energy than a single antenna. Additionally, the CBTC algorithm only supports minimum connectivity and therefore, any node failure may partition the network. A recent CBTC algorithm [3] constructs a k-connected topology if $\alpha = 2\pi/3k$. In such networks, for each $p < k$, failure of p nodes does not disconnect the network.

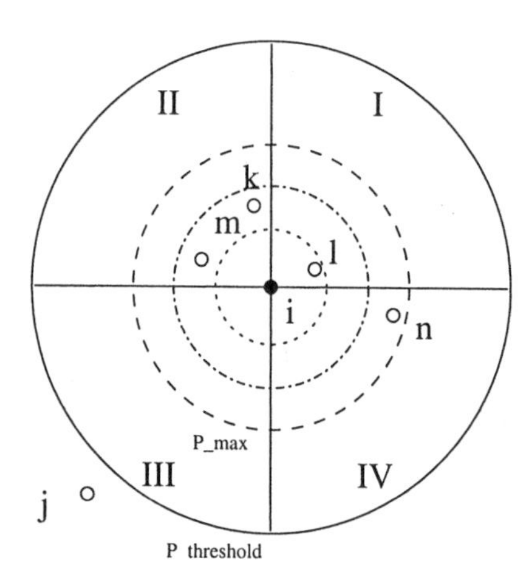

Figure 6.5. Neighbor discovery in the cone-based algorithm, $\alpha = \frac{\pi}{2}$. Node i adjusts its power level to P_{max} to reach all neighbors in all cones. Although, cone III (due to node j being outside the $P_{threshold}$ range), node i does not unnecessarily adjust P_{max} to $P_{threshold}$.

Local Minimum Spanning Tree. The final approach [40] uses purely local information to build a minimum energy spanning tree of the network. Connectivity is only maintained between two nodes if the link between them is part of the spanning tree. The Local Minimum Spanning Tree (LMST) algorithm is composed of three phases: *Information collection*, *Topology construction* and *Transmit power level determination.*

For the information collection phase, nodes determine their local topology, where local is defined by reachability at the threshold power level. All nodes periodically announce their location by broadcasting HELLO messages at the threshold power level. These HELLO messages are used to define the graph $G(V, E)$, where V is the set of all nodes, E is the set of links, and a link exists between two nodes if they can reach each other at the threshold power level. Each node collects the HELLO messages to determine its *visible neighborhood*, $NV(G)$, where $NV_i(G)$ is defined as the set of nodes that node i received a HELLO message from. Locally, node i maintains the graph $G_i = (V_i, E_i)$, where G_i is the induced subgraph of $G(V, E)$ such that $V_i = NV_i$ and E_i is the set of all links in G with both endpoints in V_i.

In the topology construction phase, each node i builds a local minimum spanning tree for its visible neighborhood using Prim's algorithm [9]. Specifically, a power efficient minimum spanning tree is built using G_i as the base graph. The weight of each edge is assigned to be the distance between the nodes. Although the weight of an edge in G_i should ideally be the power level required between the nodes, the weight can be approximated as the distance between the nodes since power consumption is an increasing function of distance. At the end of the topology construction phase, node i selects node j as its neighbor if the link to node j is is part of the minimum spanning tree. Finally, each node determines the specific power level needed to reach all of its neighbors by measuring the receive power of the periodically broadcast HELLO messages.

The result of running the LMST algorithm is a directed graph G_0, which may contain unidirectional links if two nodes do not both select each other as neighbors. Figure 6.6 illustrates an example where the topology derived using LMST contains such unidirectional links. There are 6 nodes in $V = \{i, j, k, l, m\}$, where $d(i, j) = d < d_{thresh}$ and $d(i, n) < d_{thresh}$. Nodes k, l and m are outside the threshold transmission range of node i. Therefore, $NV_i = \{j, n\}$. On the other hand, all nodes are in the threshold transmission range of node j and so, $NV_j = \{i, k, l, m, n\}$. Node i maintains links to both nodes j and m as its neighbors since both of these links are part of its local minimum spanning tree (see the solid lines in Figure 6.6). However, node j only keeps node k as its neighbor based on its local minimum spanning tree (see the dashed lines in Figure 6.6). Therefore, the link between node i and node j is unidirectional. However, there exists a route from node j to node i though other nodes.

If the underlying network topology, G, is connected, the unidirectional topology found by the LMST algorithm, G_0, is also connected. In G_0, either two nodes i and j are directly connected, as in G, or there is a minimum energy route from node i to node j. Therefore, G_0 is strongly connected (i.e., it is guaranteed that there exists a route from every node i to every node j). To eliminate the need to deal with unidirectional links, which break some existing routing protocols, a bidirectional topology can be constructed by either deleting all unidirectional edges, G_0^-, or adding reverse edges where unidirectional edges exist, G_0^+ (see Figure 6.7). Both G_0^- and G_0^+ preserve the connectivity of G_0. Since all links in G_0 also exist in G_0^+, it follows that G_0^+ preserves the connectivity of G_0. Similarly, the removal of unidirectional links does not affect the existence of a route between any two nodes. Since each node uses its own local minimum spanning tree to determine its neighbors, a unidirectional link can exist between node i and node j if node j found a more energy efficient route to node i through other nodes in its own visible neighborhood. Since all links are assumed to be symmetric, removal of the unidirectional link simply forces node i to use the energy-efficient route found by node j, maintaining the connectivity in the network. However, there exists a tradeoff between the two choices. While G_0^- is a simpler topology and is more efficient in terms of spatial reuse, G_0^+ provides more routing redundancy.

The LMST algorithm constructs a connected network topology using only local information. On the other hand, each node must be equipped with the ability to gather its location information. Another limitation is that the channel propagation model assumes symmetric channel conditions at both ends of the

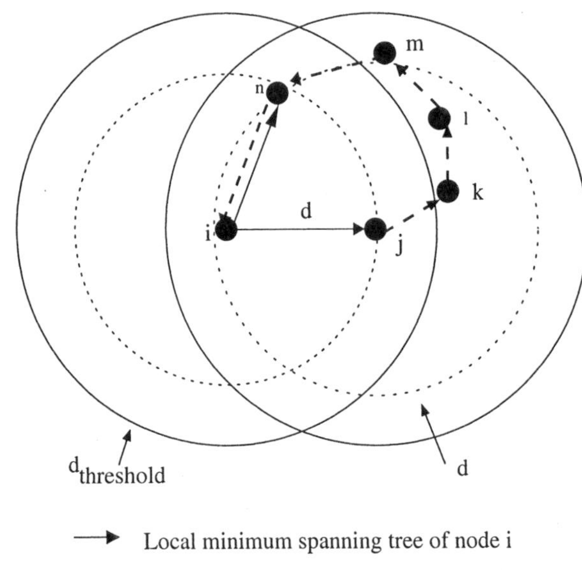

Figure 6.6. An example of unidirectional links using LMST. There are 6 nodes, $V = \{i, j, k, l, m, n\}$. The visible Neighborhood of node i is $NV_i = \{j, n\}$ and the neighbors of node i are nodes j and n. The visible Neighborhood of node j is $NV_j = \{i, k, l, m, n\}$ and only node k is its neighbor. Therefore, $i \rightarrow j$ but $j \nrightarrow i$.

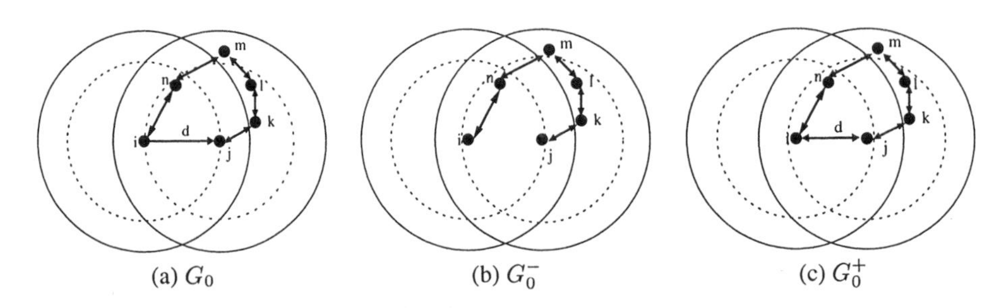

Figure 6.7. Example topologies created by the LMST algorithm

link. Given this assumption, the power level to reach a neighbor can be determined from the receive power of the HELLO messages from the neighbor. However, in practice, the noise levels at different nodes results in asymmetric conditions, limiting the effectiveness of this model.

Mobility. Since most if not all of the nodes in an ad hoc network are expected to be mobile, the topology is expected to change dynamically, implying that a new minimum energy topology must be found. The impact of node movement on a network using minimum energy topology control can be captured by looking at the movement of a single node. If the node moves closer to other nodes, communication can still be supported. However, if the node movement results in a smaller neighborhood for a node i (i.e., node i could now use a lower power level to reach all nodes in its neighborhood), node i may not know about this change and continue using an unnecessarily high power level. If the node moves away (i.e., outside of the current range of the nodes it is communicating with), the network may be partitioned. All of the protocols discussed in this section find minimum energy topologies for a given graph defined by the location of nodes in the ad hoc network. In this section, we discuss how each of the protocols deals with mobility.

COMPOW [45] should recompute the common power for the network each time a node moves to support energy efficient communication and to avoid partitions. To avoid having to determine when these changes occur, COMPOW relies on routing updates generated by the proactive routing protocol to learn about such changes and to determine a new common power. However, proactive protocols are known for their poor performance and lack of convergence in the presence of mobility and therefore, COMPOW can only handle limited mobility.

For the Connected MinMax approach [50] two distributed heuristics are introduced to support mobility. In LINT (Local Information No Topology), each node is configured with three parameters: the desired node degree, d_d, the maximum node degree, d_h and the minimum node degree, d_l. Each node periodically checks its active neighbors and adjusts its power level to stay within these thresholds. In particular, the node reduces its power level if the degree is higher than d_h and increases its power level if the degree is lower than d_l. The magnitude of the power change is a function of d_d and the current degree (i.e., the further apart the current degree and the desired degree, the higher the power change). A significant limitation of LINT is that it may not provide a connected network. LILT (Local Information Link-State Topology) tries to address this problem by exploiting global topology information available from routing protocols. Initially, all nodes transmit with the threshold power level, which results in a maximally connected network. After this initialization, power levels are adjusted based on the desired node degrees, similar to LINT. Additionally, if nodes detect a disconnection in the network via route updates, they increase their power levels to the threshold power level again. However, LILT, similar to LINT, cannot guarantee network connectivity during convergence, especially in a highly mobile environment.

In the Enclosure algorithm [52], each node periodically re-computes its enclosure to find the enclosure graph of the network. The frequency of enclosure computations should be chosen to be frequent enough to accommodate energy cost changes. However, if enclosures are computed too often, unnecessary energy may be consumed. The chosen frequency of enclosure updates must address this trade-off for energy efficient operation of the network. However, the choice of an appropriate update frequency is not addressed in [52].

To deal with mobility, CBTC uses a simple neighbor discovery protocol: each node uses beaconing messages (i.e., HELLO messages) to announce that it is still alive. The beacon includes the node ID and the power level of the beacon. A neighbor is considered to have moved away (or failed) if no beacons are received from this neighbor within a certain time interval, T. Each node reconfigures its neighborhood if there are any α-gaps (i.e., at least one of the α-degree cones is empty) or as new nodes are discovered. However, network connectivity is not guaranteed in the presence of frequent topology changes. Additionally, the choice of the time interval for HELLO messages and the time interval T is not addressed in [39].

LMST [40] must rebuild the local minimum spanning trees in the presence of mobility. To this end, the interval between two information exchanges (i.e., two HELLO messages) is determined by a probabilistic model. Based on the knowledge of the number of nodes in the network and the maximum node speed, a node computes the probability that a new node joins its neighborhood or that a neighboring node leaves its neighborhood. These two probabilities define the probability that the visible neighborhood of a node changes. A threshold update interval can be chosen to accommodate the expected changes. However, due to its probabilistic nature, LMST may not guarantee connectivity at all times.

In summary, these topology control protocols can only deal with limited mobility and do not guarantee connectivity in the presence of high mobility in the network.

6.2.3 Energy-Aware Routing

Routing protocols for ad hoc networks generally use hop count as the routing metric, which does not necessarily minimize the energy to route a packet [16]. Energy-aware routing addresses this problem by finding energy-efficient routes for communication. At the network layer, routing algorithms should select routes that minimize the total power needed to forward packets through the network, so-called *minimum energy routing*. However, minimum energy routing may not be optimal from the point of view of network lifetime and long-term connectivity, leading to energy depletion of nodes along frequently used routes and causing network partitions. Therefore, routing algorithms should evenly distribute forwarding duties among nodes to prevent any one node from being overused (i.e., *capacity-aware routing*). Hybrid protocols explore the combination of minimum energy routing and capacity-aware routing to achieve energy efficient communication while maintaining network lifetime.

Minimum Energy Routing. The routing metric used by minimum energy routing is the per-hop minimum power level $P(i, j)$ needed for node i to reach node j. The total power level for route r, P_r, is the sum of all power levels $P(i,j)$ along the route:

$$P_r = \sum_{i=0}^{D-1} P(n_i, n_{i+1}), \tag{2.6}$$

where nodes n_0 and n_D are the source and destination, respectively.

Minimum total transmission power routing (MTPR) [54] [15] [60] [61] [58] finds a minimal power route s such that:

$$P_s = \min_{r \in A} P_r, \tag{2.7}$$

where A is the set of all possible routes. Based on a given minimum energy topology that defines the maximum power level for all nodes, MTPR finds the minimum energy routes optimizing the power level for each hop. In contrast, PARO [19] is a minimum energy routing protocol ad hoc networks that discovers minimum energy routes on demand. PARO assumes that all nodes are located within direct transmission range of each other and that a source node initially uses the threshold power level to reach the destination. Each node capable of receiving the packet determines if it should intervene and forward the packet to the destination itself to reduce the energy needed to transmit the packet. Although, PARO is designed for one-hop ad hoc networks, the optimization can be used by any pair of communicating nodes, which allows extending PARO to multi-hop networks.

Given this definition of minimal power routing, both MTPR and PARO favor routes with more hops (i.e., more shorter hops vs. fewer longer hops). Since the power level, and so the transmission energy consumption, depends on distance (proportional to d^n), the energy consumed using many short hops may be less than the energy consumed using fewer longer hops [19] [41] [58] [60]. However, the more nodes involved in routing, the greater the end-to-end delay. Additionally, a route consisting of more hops is likely to be unstable due to the higher probability of the movement or failure of intermediate nodes. Furthermore, both protocols ignore the energy consumed at the relay nodes to receive the packets. Based on these observations, the routes found by MTPR and PARO may not be efficient. To overcome these problems, the energy consumed when receiving the packet should be included into the routing metric [64] [52], which is likely to result in the use of shorter routes. An even more accurate metric should include the total energy consumed in reliably delivering the message to its destination (e.g., the energy cost of link-layer retransmissions) [4]. In particular, it is essential to avoid links with relatively high error rates to reduce the energy consumed to reliably transmit packets.

Capacity-Aware Routing. Assuming all nodes in the network are equally important, no node should be used for routing more often than other nodes.

However, if many minimum energy routes all go though a specific node, the battery of this node is drained quickly and eventually the node dies. Therefore, the remaining battery capacity of a node should be used to define a routing metric that captures the expected lifetime of a node, and so, the lifetime of the network.

Given c_i^t, the battery capacity of node i at time t, the function $f_i(c_i^t)$ captures the cost to forward packets for a node i. This cost can be defined as the inverse of the remaining battery capacity and modeled as [58] [60]:

$$f_i(c_i^t) = \frac{1}{c_i^t}. \tag{2.8}$$

The battery cost metric for route r at time t, R_r, can then be determined as:

$$R_r = \max_{i \in r} f_i(c_i^t). \tag{2.9}$$

Therefore, the desired capacity-aware route s, where A is the set of all possible routes satisfies:

$$R_s = \min\{R_r | r \in A\}. \tag{2.10}$$

It must be noted that the choice of $f_i(c_i^t) = \frac{1}{c_i^t}$ does not consider the effect of the traffic load on the node battery capacity. To this end, *drain rate* is proposed as a metric to measure the *energy dissipation rate* at a given node [33]. The *Minimum Drain Rate (MDR)* algorithm determines the battery cost metric of route r, R_r, as:

$$R_r = \min_{i \in r} \frac{c_i^t}{DrainRate_i}, \tag{2.11}$$

and capacity-aware route s satisfies:

$$R_s = \max\{R_r | r \in A\}. \tag{2.12}$$

Incorporating the battery cost into the routing protocol prevents a node from being overused. However, there is no guarantee that minimum energy routes are found by the routing protocol. Therefore, capacity-aware routing may consume more energy to route traffic, which can reduce the lifetime of the network.

Hybrid Solutions (Minimum Energy/Maximum Capacity). Hybrid solutions try to find minimum energy routes while maximizing the lifetime of the network. To this end, *Conditional Max-Min Battery Capacity Routing (CMM-BCR)* [64] follows minimum energy routing as long as some routes between the source and the destination have sufficient remaining battery capacity (i.e., above a certain threshold). The battery capacity of a route r, R_r^c, is:

$$R_r^c = \min_{i \in r} c_i^t, \tag{2.13}$$

and minimum energy routing is followed as long as:

$$R_r^c \geq \delta \, for \, any \, r \in A. \tag{2.14}$$

If all routes are below the energy threshold δ, capacity-aware routing is used to determine the route to choose. The benefit of such an approach comes from the fact that capacity-aware routing is only used when critical nodes in the network have low battery levels. The efficiency of the CMMBCR depends on the energy threshold δ. However, it is not straightforward how to determine δ.

The *Conditional Minimum Drain Rate (CMDR)* protocol [33] limits route choices for MTPR to routes only containing nodes with a lifetime higher than a given threshold (i.e., $\frac{c_i^t}{DrainRate_i} \geq \gamma$). If no such route exists, CMDR switches to the MDR scheme. To overcome the difficulty of selecting a value for δ in CMMBCR, CMDR uses γ, which is an absolute time value based on the current traffic conditions.

The *max-min* $z \cdot P_{min}$ algorithm [41] minimizes energy consumption and maximizes the minimum residual energy of the nodes. If the minimum energy route has energy consumption P_{min}, routes with higher minimum residual energy can be used as long as the energy consumption is less than $z \cdot P_{min}$. The z-factor, similar to CMDR, is computed based on the minimum lifetime of the nodes.

All three of the above algorithms find minimum energy routes when nodes have sufficient residual energy and switch to capacity-aware routing as the battery capacity of the nodes decreases or the lifetime decreases beyond a predefined threshold. In contrast, the cost metric of a link (i, j) can be chosen to represent both the transmission power cost of the link and the initial and residual energy of node i [7] [60]. Specifically, link cost, $cost_{i,j}$, can be computed as [7]:

$$cost_{i,j} = e_{ij}^{\alpha}(c_i^t)^{-\beta}(c_i^0)^{\theta}, \tag{2.15}$$

where e_{ij} is the energy used to transmit and receive on the link, c_i^t is the current capacity of node i, c_i^0 is the initial capacity of node i and α, β and θ are non-negative weights. The link cost function computed in this fashion emphasizes the energy expenditure term when nodes have high battery capacity. As the residual energy of the nodes decreases, the battery capacity term is more emphasized.

To avoid depletion of nodes along common minimum energy routes, another approach is to occasionally use sub-optimal routes [55]. Basically, possible routes between a source and destination are used with a probability based on the energy metric in Equation 2.15.

6.3 Idle-time Energy Conservation

Effective idle-time energy conservation necessarily spans all layers of the communication protocol stack. Each layer has access to different types of information about the communication in the network, and thus, uses different mechanisms to support energy conservation. MAC layer protocols can save energy by suspending the communication device during short-term idle periods in communication (i.e., operate in a power-save mode). Such fine-grained control requires integrated knowledge of transitions between device suspend and resume in the MAC protocol to insure the communicating nodes are both awake. The delay overhead from waking up a suspended device can negatively impact communication in the network and so power-save modes should not always be used. Power management protocols integrate global information based on topology or traffic characteristics to determine transitions between active mode (i.e., never suspend) and power-save mode.

6.3.1 Communication Device Suspension

When not transmitting, a wireless communication device is continuously listening for incoming transmissions. This listening cost can be quite high since a node must try to receive a packet to see if there is actually a packet being transmitted to it or any other node. If there are currently no transmissions destined for a given node, this listening wastes significant amounts of energy [18] [35]. In wireless communication devices, the cost of listening is only slightly lower than the actual cost of receiving, since listening requires minimal processing overhead compared to receiving [13]. Table 6.1 only lists receive costs since most specifications do no include idle costs. However, measurements show a significant difference between idle and receive costs, depending on the specific device [18].

Listening costs can be reduced by shutting off the device or placing the device in a low-power state when there is no active communication [26] [35]. The low-power state turns off the receiver inside the device, essentially placing the device in a suspended state from which it can be resumed relatively quickly. In general, the suspend costs for most current devices are low enough that the overhead from staying in a suspended state is minimal. In a completely off state, the device consumes no energy. However, the time it takes to resume a device from a completely turned off state can be prohibitively long (i.e., on the order of hundreds of milliseconds) and may even consume extra energy to re-initialize the card. The choice about whether to use suspension or whether to turn off the device must include information about the expected communication patterns for a node. Given that all nodes in an ad hoc network participate in routing and forwarding, we mainly focus on suspending the device.

The goal of any device suspension protocol is to only remain awake when there is active communication for a node and otherwise suspend. In general, active communication is defined to be communication, unicast, multicast or broadcast, that originates from or is destined to a node. However, many ad hoc routing protocols take advantage of the fact that all communication in an ad hoc network is inherently broadcast and snoop on communication in their neighborhood to populate their routing tables. Allowing a node to suspend its device limits the node's ability to snoop on communication between neighboring nodes. To date, there has been little evaluation of the impact of device suspension on route caching.

When a communication device is suspended, the node is effectively cut off from the rest of the network. While it is relatively simple to resume the device when the node has packets to send, the challenge comes from dealing with packets destined to a node with a suspended communication device. Two major problems arise when the destination device is suspended. First, sending nodes must know when the receiving node's communication device is suspended. If the receiver is active, the sender should transmit immediately. If the receiver is suspended, the sender should buffer the packet and wait until the receiver wakes up to transmit it. Packets destined to a suspended device experience delay on the order of the length of time the device spends suspended before it resumes to check for pending transmissions. If the receiver is suspended for too long, the sender's buffers could fill up, eventually causing the packet to be dropped. If the sender thinks a suspended receiver is active, the sender tries to transmit the packets with no success, unnecessarily wasting its own energy and potentially dropping the packets because it thinks the receiver is not in range. Second, the suspended device can only guess when there are packets destined for the node. If it resumes when there are none, it again wastes energy listening to an idle channel. If it waits too long to resume, pending packets are unnecessarily delayed or even dropped.

Since both the sender and receiver must be awake to transmit and receive, it is necessary to ensure an overlap between awake times for nodes with pending communication. We discuss two types of protocols used to manage the suspend/resume cycles for individual nodes: *periodic resume* and *triggered resume*.

Periodic Resume. A simple technique for managing the suspend/resume cycles for nodes is to allow the nodes to suspend most of the time and to resume periodically to check for pending transmissions. If no packets are destined for a node when it checks, the node can again suspend its device. If a node has some packets destined for it, it remains awake until there are no more packets or until the end of the cycle. Nodes can be notified of pending transmissions through an out-of-band control channel (i.e., part of the channel is reserved for

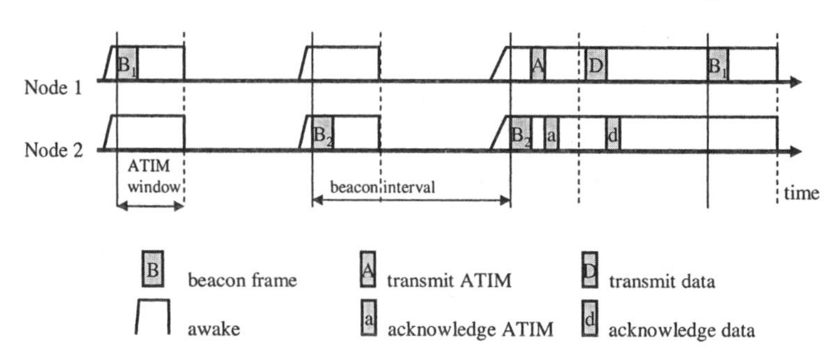

Figure 6.8. IEEE 802.11 Power-save Mode

control messages) [26] [17] [65] or through in-band signaling [71]. The success of a periodic resume approach is to ensure that nodes that want to communicate with each other are awake at the same time to coordinate the notification and ensure that the receiver remains awake to receive the pending transmission. We discuss two approaches to periodic resume: *synchronous* and *asynchronous*.

Synchronous. If the periods of all nodes in the ad hoc network are synchronized, all nodes are guaranteed to have overlapping awake times, and so can easily have overlapping out-of-band channels for notification. Such a synchronized solution is specified in IEEE 802.11 Power Save Mode (PSM) [26], which also provides low-level support for buffering packets for sleeping nodes and synchronizing nodes. In power-save mode, all nodes in the network are synchronized to wake up at the beginning of a beacon interval (see Figure 6.8). To maintain synchronization, beacon messages are sent at the beginning of every beacon interval. Synchronization in single-hop networks only requires the transmission of one beacon per interval. To determine if and when to send a beacon, nodes use a random backoff algorithm. Any node that hears another node's beacon before it sends its own cancels its beacon and synchronizes to that beacon. This algorithm has been proposed for use in multiple-hop wireless networks. While there is the possibility that the algorithm will not converge in dynamic environments, the randomness of the algorithm should enable convergence in many environments.

Broadcast, multicast or unicast packets to a power-saving node are first announced during the period when all nodes are awake. The announcement is done via an ad hoc traffic indication message (ATIM) inside a small interval at the beginning of the beacon interval called the ATIM window. Channel access and contention resolution for communication during the ATIM window follow the same rules as during normal communication. A node that receives a directed ATIM during the ATIM window (i.e., it is the designated receiver) sends an acknowledgement and stays awake for the entire beacon interval waiting for the packet to be transmitted. A node with a pending ATIM that overhears an ATIM acknowledgement from that node need not send another indication since it already knows that the node will remain awake to receive. Broadcast/multicast packets announced in the ATIM window are not directed to a specific node and so are not acknowledged. However, both broadcast and multicast indication messages cause all nodes (or just the nodes in the multicast group) to stay awake for the entire beacon interval. Since the neighbor sets of two nodes is not likely to be exactly the same, nodes send broadcast/multicast indications even if they have already heard one during the current ATIM window.

Immediately after the ATIM window, nodes can transmit buffered broadcast, multicast or unicast packets addressed to nodes that are known to be awake (e.g., nodes that have acknowledged a previously transmitted ATIM). Following the transmission of all announced packets, nodes can continue to transmit packets destined to nodes that are known to be awake for the current beacon interval. The state of a node (i.e., awake or suspended), can be determined by snooping ATIM acknowledgements or by snooping control messages during active communication.

Figure 6.8 shows the interactions between two nodes using IEEE 802.11 PSM in an ad hoc network. During the first two beacon intervals, no packets are pending for either node. The two nodes randomly send beacon messages to maintain synchronization. In the third interval, node 1 has a packet to send to node 2 and so sends a directed ATIM, which is acknowledged by node 2. After the ATIM window has ended, node 1 knows that node 2 is awake and sends the packet using normal channel access rules.

In protocols like IEEE 802.11 PSM that use an out-of-band channel to announce pending transmissions, the throughput of the network is limited to the amount of data that can be announced in the channel. Essentially, if a node cannot send an indication message to wake up its destination, it must buffer its packets until the next beacon interval. If this continues to happen, the node's buffer eventually fills up and packets are dropped.

Asynchronous. Given the expected mobility of nodes in ad hoc networks, clock synchronization may be difficult to maintain. If the clocks are not well synchronized, nodes may not be awake to hear each other's notification messages. However, it is possible to allow nodes to use asynchronous cycles if there is a guarantee that communicating nodes' awake times overlap [17] [28] [65] [71]. The basic idea behind such *asynchronous wake up protocols* is that nodes stay awake longer so there is a guaranteed chance to listen for pending communication from other nodes. Asynchronous approaches can be used with and without notification messages (e.g., ATIMs in IEEE 802.11). All approaches

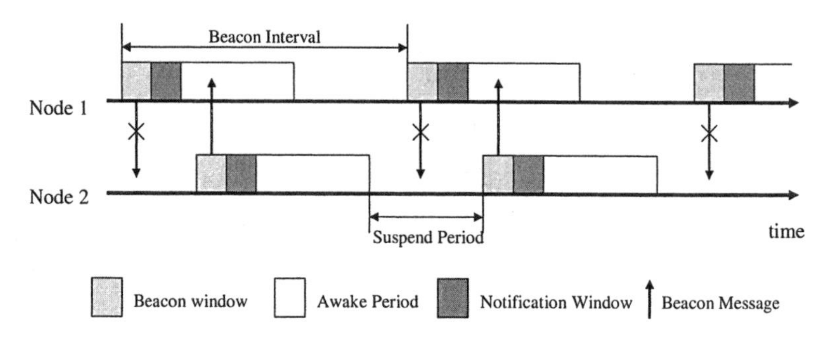

Figure 6.9. Mis-matched Beacon Intervals. Node 2 can never hear the ATIM from node 1.

discussed in this section use beacon messages to inform listening nodes of the beaconing node's presence and of the start of its awake period.

If notification messages are used, the notification window (e.g., the ATIM window in IEEE 802.11 PSM) of the transmitting node must overlap with the awake period of its neighbor node for which it has a packet to transmit. In these approaches [17] [28] [65] [71], each interval is divided into an awake period and a suspend period. Beacon and notification messages are still sent at the beginning of every awake period. To guarantee the overlapping of notification windows and awake periods for nodes with pending communication, awake periods must be at least half of the beacon interval. In other words, every node is awake at least half of the time. However, this change alone does not guarantee overlap. For example, in Figure 6.9, node 2 always misses node 1's beacons. This problem can be fixed by either having the notification window be at the beginning of even periods and at the end of odd periods [28] [65] (see Figure 6.10), or by having two notification windows, one at the beginning of a period and one at the end [17] (see Figure 6.11). Both approaches ensure that at least every other notification window overlaps with a neighbor's awake period. However, requiring a node to remain awake at least half of the time limits the amount of energy that can be saved by these approaches.

The amount of awake time can be reduced in one of three ways. First, a node can remain fully awake once every T beacon intervals [28] (see Figure 6.12). This approach reduces the amount of time a node must remain awake, but increases the delay to transmit to a suspended node. A message could be delayed up to T times the length of the beacon interval before the node can receive a notification message.

The second approach improves on the first by increasing the number of beacon intervals in the cycle but also increasing the number of fully awake intervals [28] [65]. Additionally, the number of beacon messages is reduced by only requiring beacon messages during awake intervals. Essentially, each n^2 intervals, a node stays fully awake $2n-1$ intervals. These $2n-1$ intervals must

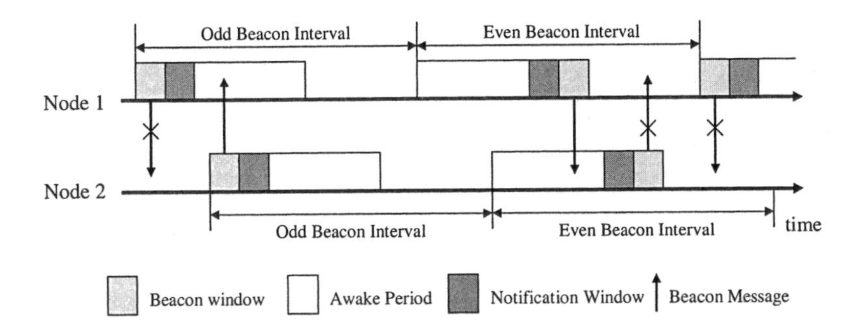

Figure 6.10. Alternating odd and even cycles ensure that all nodes can hear each other's notification messages.

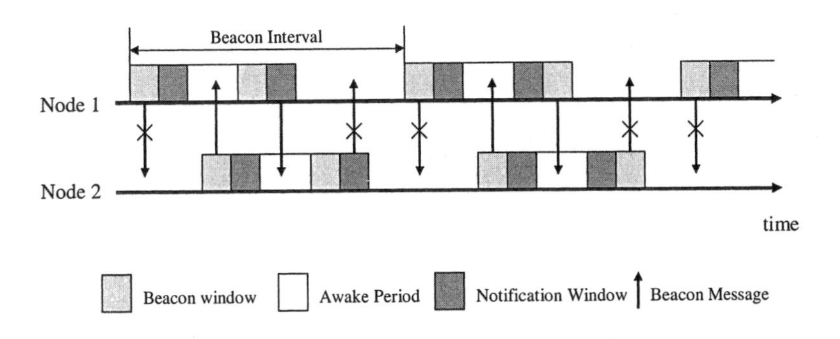

Figure 6.11. Using two notification windows guarantees overlap.

form a quorum, ensuring a non-empty overlap set between any two neighbors. If the n^2 intervals are arranged as a 2-dimensional $n \times n$ array, each host can pick one row and one column of entries as awake intervals (i.e., $2n - 1$) (see Figure 6.13). No matter which row and column are chosen, two nodes are guaranteed to have at least two overlapping awake intervals, guaranteeing the chance to hear each other's notification messages. For example, if $n = 4$, node i chooses row 0 and column 1 and node j chooses row 2 and column 2, they both stay awake during intervals 2 and 9 (see Figure 6.14). This approach improves the average delay to wake up a node since nodes are guaranteed at least two overlapping awake intervals per cycle. However, in the worst case, the overlapping intervals could be right next to each other, resulting in a potential delay up to the length of the whole cycle.

The third approach eliminates the need for notification messages, although still requires beacon messages during awake periods. In this approach, each nodes cycles through a pattern of awake and suspend periods [71]. Every node uses the same pattern, although they may be offset from each other in time. Any pattern of any length can be used as long as it guarantees sufficient overlapping awake intervals between any two nodes. If the number of overlapping intervals is 1, a feasible pattern can be found if the cycle length is a power of a prime number. Other cycle lengths require more overlapping slots. For example, consider a cycle of seven slots to achieve one overlapping slot per pair of nodes. Figure 6.15 shows seven nodes, each with the same pattern, but offset from each other by one slot. This pattern of (awake, awake, suspend, awake, suspend, suspend, suspend) guarantees that every node has at least one overlapping awake interval with every other node, ensuring that each pair of nodes has the opportunity to communicate at least once per cycle. The synchronization between nodes is not required for correctness. We can see in Figure 6.16 that if the nodes' slots are not synchronized, they are still guaranteed to hear each other's beacon messages once per cycle. If one slot is not sufficient to transmit all pending packets, the receiving node listens for the in-band signals in an augmented MAC layer header and remains awake during the next slot to receive the remaining buffered packets. The delay imposed by this approach depends on the number of overlapping awake intervals per cycle.

While asynchronous wake up removes any overhead from maintaining synchronization in the network, a node may spend significantly more time awake than in a synchronous approach. Additionally, all current approaches incur more delay than a synchronous approach. One major drawback of asynchronous wake up is that broadcast support is only provided if the awake periods of all nodes within transmission range of the sender overlap. One approach to solving this problem is to transmit the broadcast message multiple times. However, it is unclear what impact this will have on total energy consumption or on communication in the network. Routing protocols are a particular concern since

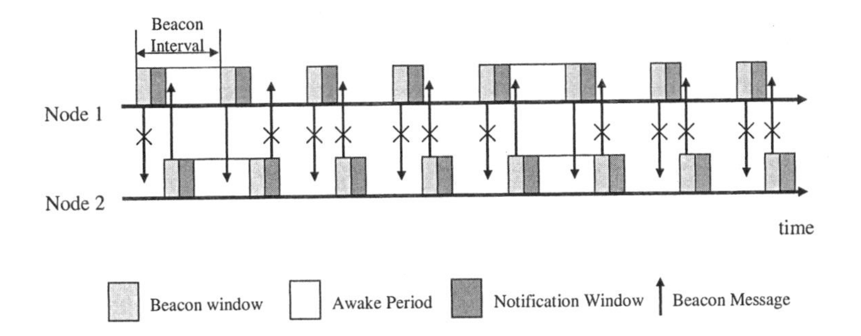

Figure 6.12. Nodes remain awake once every T intervals ($T = 4$). However, communication is delayed up to T times the length of the beacon interval

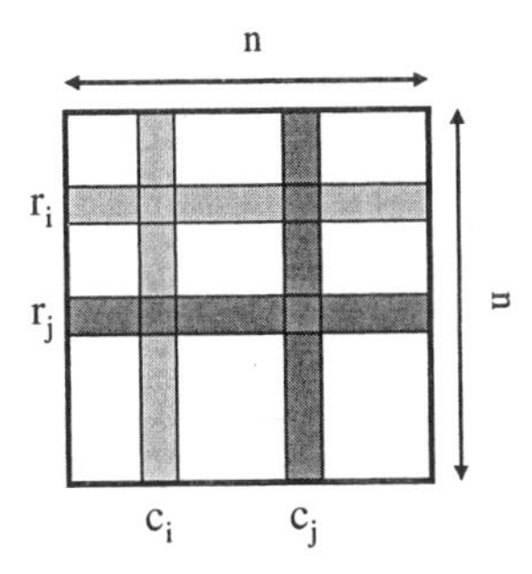

Figure 6.13. Nodes remain awake once $2n - 1$ every n^2 intervals. Nodes each choose one row and one column (i.e., node i chooses row r_i and column c_i and node j chooses row r_j and c

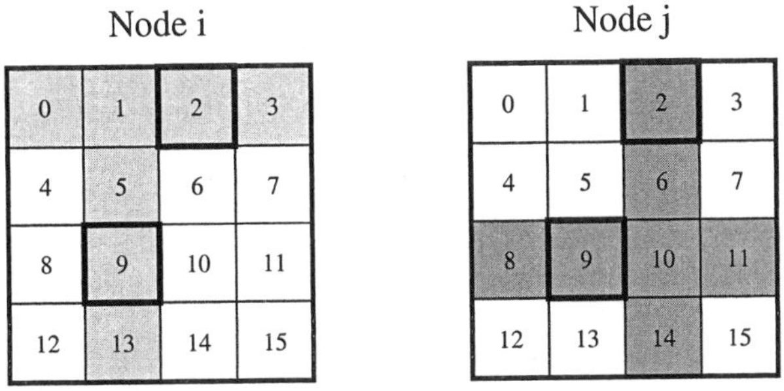

Node i

0	1	2	3
4	5	6	7
8	9	10	11
12	13	14	15

Node j

0	1	2	3
4	5	6	7
8	9	10	11
12	13	14	15

Figure 6.14. Node i chooses row 0 and column 1 and node j chooses row 2 and column 2. Both stay awake during intervals 2 and 9 ($n = 4$).

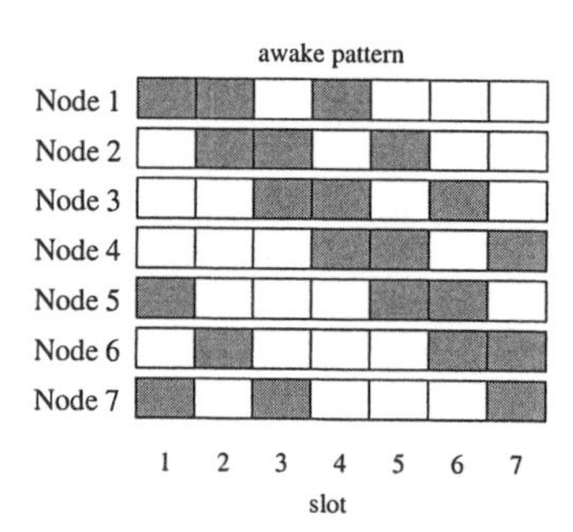

Figure 6.15. Slot allocations determine when each node remains awake. This figure shows an example slot allocation that guarantees at least one overlapping slot between any two nodes.

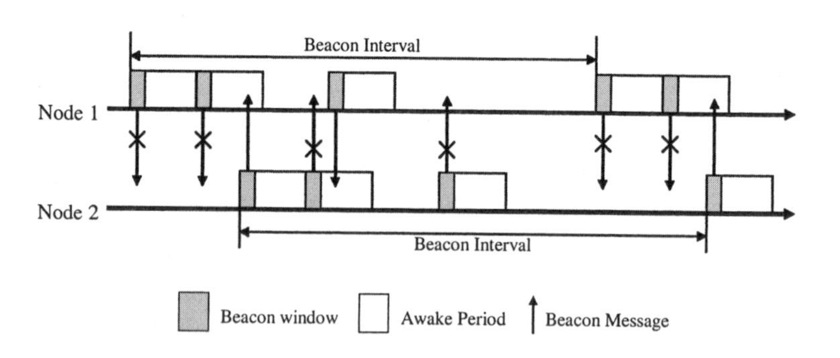

Figure 6.16. Nodes with offset slots are guaranteed to hear each other's beacon messages at least once per cycle..

they typically discover and maintain routes by broadcasting requests through the network.

Triggered Resume. To avoid the need for periodic suspend/resume cycles, a second control channel can be used to tell the receiving node when to wake up, while the main channel is used to transmit the message [1] [49] [53] [56] [57]. To be effective, the control channel must consume less energy than the main channel and also must not interfere with the main channel. For example, transmitting in the 915Mhz [49] [56] or using RFID technology [1] does not interfere with IEEE 802.11, and both consume significantly less energy.

RTS [57] or beacon messages [53] [56] are sent using the control channel to wake up intended receivers, which first respond in the control channel and then turn on their main channel to receive the packet. After the packet transmission has ended, the node turns its radio off in the main channel. Similar to IEEE 802.11, sleeping nodes with traffic destined for them are woken up. However, the decisions about when a node should go back to sleep can be based on local information. The out-of-band signaling used by triggered resume protocols avoids the extra awake time needed by asynchronous periodic resume protocols. Triggered resume protocols like PAMAS [57] and Wake-on-Wireless [56] assume that the radio in the control channel is always active, avoiding the clock synchronization needed by synchronous periodic resume protocols such as IEEE 802.11. Additional savings can be achieved on the control channel using any of the periodic resume approaches. For example, STEM [53] uses a synchronized periodic resume protocol, saving energy in the control channel at the cost of requiring node synchronization.

Triggered resume protocols do not provide mechanisms for indicating the power management state of a node, and so senders assume a receiver is suspended by default. Essentially, the power management state is only maintained on a per-link basis between nodes with active communication. Therefore, it is possible that a sending node experiences the delay from waking up a receiver node, even if the receiver is already awake due to recent communication with a third node.

The limitations of triggered resume protocols come from the complexity of requiring two radios on one node. First, two radios are certainly more expensive than one. Although, if dual radio approaches become popular, the extra cost could become less significant. Second, the characteristics of the wireless communication channel of the two radios can differ significantly in terms of transmission range and tolerance to interference. There is no guarantee that the main channel is usable even if the control radio can successfully transmit to the receiver, causing the receiving node to resume and the sending node to try to transmit needlessly. Similarly, a usable main channel is not accessible if the control channel is not usable, needlessly preventing communication from occurring.

6.3.2 Power Management

In ad hoc networks, suspending a node's communication device can impact communication at multiple layers of the protocol stack. At the MAC layer, uncoordinated suspension between two nodes can prevent the nodes from communicating. At the routing layer, a node that is suspended could be miscategorized as having moved away and so cause a route to break, incurring unnecessary route recovery overhead. Additionally, current device suspension protocols place limitation on the amount of data that can be supported in the network.

If the coordination of suspend and resume states between communicating nodes causes too many packets to be dropped or delayed, the suspension of devices can actually end up consuming more energy [2] [34] [72]. Similarly, if not enough data can be supported in the network, the suspension of devices can limit the effectiveness of the network. Communication in the network can be improved by allowing higher layer decisions about if a device should ever use power-saving techniques. In this context, a node can be in one of two power management modes: active mode and power-save mode. In active mode, a node is awake and may receive at any time. In power-save mode, a node is suspended most of the time and resumes periodically to check for pending transmissions, as described in the previous section. The role of a power management protocol is to determine when a node should transition between active mode and power-save mode.

Packets traversing an ad hoc network can experience difficulties from power management at every hop, impacting the routing protocols and the productivity of the network [72]. The major challenge to the design of a power management protocol for ad hoc networks is that energy conservation usually comes at the cost of degraded performance such as lower throughput or longer delay. Essentially, the goal of power management is to let as many nodes use power-save mode as possible while maintaining effective communication in the network. A naive solution that only considers power savings of individual nodes may turn out to be detrimental to the operation of the whole network.

Power Management and Routing. The particular decisions about when a node should be in a power-save mode affect the discovery of routes as well as the end-to-end delay of packets. Similar to ad hoc routing protocols, power management schemes range from proactive to reactive. The extreme of proactive can be defined as always-on (i.e., all nodes are in active mode all the time) and the extreme of reactive can be defined as always-off (i.e., all nodes are in power-save mode all the time). Given the dynamic nature of ad hoc networks, there must be a balance between proactiveness, which generally provides more efficient communication, and reactiveness, which generally provides better power saving. In this space, we discuss three approaches to using power management in ad hoc networks: *reactive*, *proactive*, and *on-demand.*

Reactive Power Management. A pure power saving approach (i.e., always-off) can be considered as the most reactive approach to power management. However, a network that relies solely on MAC layer power management such as IEEE 802.11 can be highly inefficient even though some communication is still possible [72]. In an always-off network, all nodes must be woken up before any communication can occur, causing increased delay for both control (e.g., route request or route reply) and data packets. Additionally, all transmissions must be announced (e.g., via an ATIM). If the resources for announcement (e.g., the ATIM window size), cannot support the load in the network, queues fill up and packets get dropped. In a lightly loaded network, an always-off approach can generally support the traffic with little or no drops, although there is still an increased delay. However, in a heavily loaded network, the announcements become a bottleneck and little or no effective communication occurs.

Proactive Power Management. A proactive approach to power management provides some persistent maintenance of the network to support effective communication. Since routing protocols operate at the network layer, proactive power management schemes can take advantage of topological information to ensure that a specific set of nodes stays awake to provide complete connectivity for routing in the ad hoc network [5] [6] [8] [22] [67] [68]. We call this type of approach *topology management*. This differs from topology control, since topology control determines the topology for all nodes while topology management determines which nodes participate in routing in the network.

One approach to topology management is to create a connected dominating set (CDS), where all nodes are either a member of the CDS or a direct neighbor of one of the members [59] (see Figure 6.17). In general CDS-based routing, nodes in the CDS serve as the "routing backbone" and all packets are routed through the backbone. In a CDS-based power management protocol, all nodes on the CDS remain active all the time to maintain global connectivity (e.g., GAF [68] and Span [8]). All other nodes can choose to use power-save mode or even turn off completely. GAF creates a virtual grid and chooses one node in every grid location to be part of the backbone and remain awake (see Figure 6.18). All other nodes turn completely off. Span takes a slightly different approach and uses local message exchanges to allow a node to determine the effect on its neighbors if it stays awake or uses a low-power mode like IEEE 802.11 PSM. Both Span and GAF assume that sources and destinations are separated from pure forwarding nodes. In the case of mixed source/destination/forwarding nodes scenarios, the specification of both protocols is incomplete. Neither

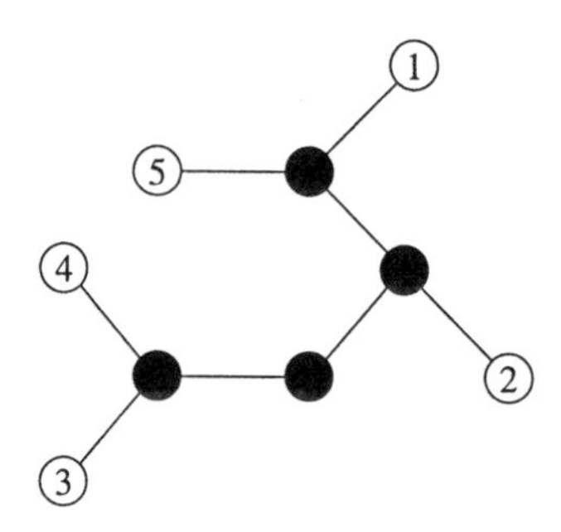

Figure 6.17. Example Connected Dominating Set. The black nodes form the CDS. Nodes 1-5 are all only one hop away from a node in the CDS.

protocol has a mechanism for signaling the data sink for incoming transmissions. In Span, it is unclear whether the election of coordinators should consider the fact that some nodes may be required to be turned on as data sources or destinations.

By taking advantage of route redundancy in dense ad hoc networks, topology management approaches save energy by turning off devices that are not required for global network connectivity. The challenge to topology management comes from the need to maintain the CDS, generally through local broadcast messages that may consume a significant amount of energy [18], especially since broadcast messages wake up all nodes for some amount of time. Additionally, the nodes chosen to participate in the CDS are periodically rotated to prevent any one node from having its battery depleted. This rotation essentially results in the formation of a new CDS, resulting in unnecessary overhead if the CDS does not change. The final limitation to these approaches comes from the fact that regardless of whether or not traffic is present in the network, all the backbone nodes must be active all the time. Essentially, even if there is no traffic in the network, some nodes are still active and consuming significant amounts of energy.

On-Demand Power Management. In response to the limitations of both reactive and proactive power management, on-demand power management eliminates the need to maintain any nodes in active mode if there is no traffic in the network by tying power management decisions to information about which nodes are used for routing in the ad hoc network [72]. In on-demand power management, all nodes are treated equal, eliminating the need to know which nodes are sources and destinations. All nodes are initially in power-save mode. Upon reception of packets, a node starts a keep-alive timer and switches to active mode. Upon expiration of the keep-alive timer, a node switches from active mode to power-save mode. The goal is to have nodes that are actively

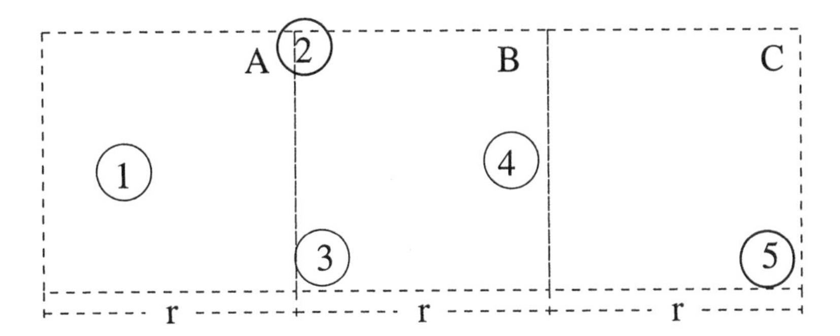

Figure 6.18. GAF's virtual grid. One node in each grid location remains awake to create a connected dominating set.

forwarding packets stay in active mode, while nodes that are not involved in packet forwarding may go into power-save mode. The key idea of on-demand power management is that transitions from power-save mode to active mode are triggered by communication events such as routing control packets or data packets and transitions from active mode to power save mode are determined by a soft-state timer.

In an ad hoc network, if a route is going to be used, the nodes along that route should be awake to not cause unnecessary delay for packet transmissions. If a route is not going to be used, the nodes should be allowed to use power-save mode. During the lifetime of the network, different packets indicate different levels of "commitment" to using a route. Knowledge of the semantics of such messages can help make better power management decisions. On one end, most control messages (e.g., link state in table-driven ad hoc routing protocols, location updates in geographical routing, route request messages in on-demand routing protocols, etc.) are flooded throughout the network and provide poor hints for the routing of data. Such control messages should not trigger a node to stay in active mode. On the other end, data packets are usually bound to a route on relatively large time scales. Therefore, data packets are a good hint for guiding power management decisions. For data packets, nodes should stay active on the order of packet inter-arrival times to ensure that no node along the route goes into power-save mode during active communication. There are also some control messages, such as route reply messages in on-demand routing protocols and query messages in sensor networks, that provide a strong indication that subsequent packets will follow this route. Therefore, such messages should trigger a node to switch to active mode. The time scale for such a transition should be on the order of the end-to-end delay from source to destination so the node does not transition back to power-save mode before the first data packet arrives.

The improvement in energy consumption comes at an increase in the initial delay of packets in a newly established route. Essentially, if all nodes along the route are asleep, they must all be woken up, incurring delay on the order of the length of the route times the time to wake up a node. However, in an active network, many nodes are expected to be awake. On-demand power management implicitly finds routes with more awake nodes, since those routes have shorter delays. Since on-demand power management favors awake nodes, it should be coupled with capacity-aware routing to support load balancing.

6.4 Conclusion

Energy conservation in ad hoc networks is a relatively new field of research. In this chapter, we have presented some of the recent proposals and specifications for achieving that goal. It is clear that there is still room for new approaches that tackle this extremely complex problem of balancing energy conservation with communication quality in dynamic ad hoc networks.

Acknowledgments

The authors wish to thank the many people who helped us bring this chapter together. We would like to specially thank Rong Zheng for her insights into energy conservation in ad hoc networks and Rob Kooper for making it all look great. Additional thanks go out to the members of the Mobius Group in the Computer Science Department at the University of Illinois, Urbana-Champaign.

References

[1] S. Agarwal, S. V. Krishnamurthy, R. H. Katz, and S. K. Dao. Distributed power control in ad-hoc wireless networks. In *IEEE Symposium on Personal Indoor and Mobile Radio Communication (PIMRC)*, 2001.

[2] M. Anand, E. B. Nightingale, and J. Flinn. Self-tuning wireless network power management. In *9th Annual International Conference on Mobile Computing and Networking (MobiCom)*, 2003.

[3] M. Bahramgiri, M. T. Hajiaghayi, and V. S. Mirrokni. Fault-tolerant and 3-dimensional topology control in wireless multi-hop networks. In *International Conference on Computer Communications and Networks (IC-CCN'02)*, 2002.

[4] S. Banerjee and A. Misra. Minimum energy paths for reliable communication in multi-hop wireless networks. In *3rd ACM International Symposium on Mobile Ad Hoc Networking and Computing (MobiHoc)*, 2002.

[5] L. Bao and J. J. Garcia-Luna-Aceves. Topology management in ad hoc networks. In *4th ACM International Symposium on Mobile Ad Hoc Networking and Computing (MobiHoc)*, 2003.

[6] A. Cerpa and D. Estrin. ASCENT: Adaptive self-configuring sensor networks topologies. In *IEEE INFOCOM*, 2002.

[7] J.-H. Chang and L. Tassiulas. Energy conserving routing in wireless adhoc networks. In *IEEE INFOCOM*, 2000.

[8] B. Chen, K. Jamieson, H. Balakrishnan, and R. Morris. Span: An energy-efficient coordination algorithm for topology maintenance in ad hoc wireless networks. In *7th Annual International Conference on Mobile Computing and Networking (MobiCom)*, 2001.

[9] T. H. Cormen, C. E. Leiserson, R. L. Rivest, and C. Stein. *Introduction to Algorithms*. MIT Press/McGraw-Hill, second edition, 2001.

[10] F. Douglis, P. Krishnan, and B. N. Bershad. Adaptive disk spindown policies for mobile computers. In *Second USENIX Symposium on Mobile and Location Independent Computing*, 1995.

[11] F. Douglis, P. Krishnan, and B. Marsh. Thwarting the power-hungry disk. In *USENIX Symposium*, 1994.

[12] R. Dube, C. D. Rais, K.-Y. Wang, and S. K. Tripathi. Signal stability based adaptive routing (SSA)for ad hoc mobile networks. *IEEE Personal Communications*, 4(1):36–45, February 1997.

[13] J.-P. Ebert, B. Burns, and A. Wolisz. A trace-based aprroach for determining the energy consumption of a WLAN network interface. In *European Wireless 2002*, 2002.

[14] T. ElBatt and A. Ephremides. Joint scheduling and power control for wireless ad-hoc networks. In *IEEE INFOCOM*, 2002.

[15] T. A. ElBatt, S. V. Krishnamurthy, D. Connors, and S. Dao. Power management for throughput enhancement in wireless ad-hoc networks. In *IEEE International Conference on Communications (ICC)*, 2000.

[16] L. M. Feeney. An energy consumption model for performance analysis of routing protocols for mobile ad hoc networks. *Mobile Networks and Applications*, 6(3):239–249, June 2001.

[17] L. M. Feeney. A QoS aware power save protocol for wireless ad hoc networks. In *Mediterranean Ad Hoc Networking Workshop (Med-Hoc-Net)*, 2002.

[18] L. M. Feeney and M. Nilsson. Investigating the energy consumption of a wireless network interface in an ad hoc networking environment. In *IEEE INFOCOM*, 2001.

[19] J. Gomez, A. T. Campbell, M. Naghshineh, and C. Bisdikian. PARO: Supporting dynamic power controlled routing in wireless ad hoc networks. *Wireless Networks*, 9(5):443–460, September 2003.

[20] P. Gupta and P. R. Kumar. *Stochastic Analysis, Control, Optimization and Applications*, chapter Critical Power for Asymptotic Connectivity in Wireless Networks, pages 547–566. Birkhauser Boston, 1998.

[21] P. Gupta and P. R. Kumar. The capacity of wireless networks. *IEEE Transactions on Information Theory*, 46(2):388–404, March 2000.

[22] W. R. Heinzelman, A. Chandrakasan, and H. Balakrishnan. Energy-efficient communication protocol for wireless microsensor networks. In *Hawaii International Conference on System Sciences (HICSS)*, 2000.

[23] Gavin Holland, Nitin Vaidya, and Paramvir Bahl. A rate-adaptive mac protocol for multi-hop wireless networks. In *7th Annual International Conference on Mobile Computing and Networking (MobiCom)*, 2001.

[24] T.-C. Hou and V. Li. Transmission range control in multihop packet radio networks. *IEEE Transactions on Communications*, 34(1):38–44, January 1986.

[25] C.-H. Hwang and A. C.-H. Wu. A predictive system shutdown for energy saving of event-driven computation. In *IEEE/ACM International Conference on Computer Aided Design*, 1997.

[26] IEEE 802 LAN/MAN Standards Committee. Wireless LAN medium access control MAC and physical layer (PHY) specifications. IEEE Standard 802.11, 1999.

[27] IEEE 802 LAN/MAN Standards Committee. Wireless LAN medium access control MAC and physical layer (PHY) specifications: Spectrum and transmit power management extensions in the 5ghz band in europe. Draft Supplement to IEEE Standard 802.11 1999 Edition, 2002.

[28] J.-R. Jiang, Y.-C. Tseng, C.-S. Hsu, and T.-H. Lai. Quorum-based asynchronous power-saving protocols for ieee 802.11 ad hoc networks. In *International Conference on Parallel Processing (ICPP)*, 2003.

[29] D. B. Johnson and D. A. Maltz. *Mobile Computing*, chapter Dynamic source routing in ad hoc wireless networks, pages 153–181. Kluwer Academic Publishers, February 1996.

[30] E.-S. Jung and N. H. Vaidya. A power control MAC protocol for ad hoc networks. In *8th Annual International Conference on Mobile Computing and Networking (MobiCom)*, 2002.

[31] V. Kawadia and P. R. Kumar. Power control and clustering in ad hoc networks. In *IEEE INFOCOM*, 2003.

[32] V. Kawadia, S. Narayanaswamy, R. Rozovsky, R. S. Sreenivas, and P. R. Kumar. Protocols for media access control and power control in wireless networks. In *IEEE Conference on Decision and Control*, 2001.

[33] D. Kim, J.J. Garcia-Luna-Aceves, K. Obraczka, J.-C. Cano, and P. Manzoni. Routing mechanisms for mobile ad hoc networks based on the energy drain rate. *IEEE Transactions on Mobile Computing*, 2(2):161–173, April-June 2003.

[34] R. Krashinsky and H. Balakrishnan. Minimizing energy for wireless web access with bounded slowdown. In *8th Annual International Conference on Mobile Computing and Networking (MobiCom)*, 2002.

[35] R. Kravets and P. Krishnan. Application-driven power management for mobile communication. *Wireless Networks*, 6(4):263–277, 2000.

[36] S.-J. Lee and M. Gerla. Dynamic load-aware routing in ad hoc networks. In *IEEE International Conference on Communications (ICC)*, 2001.

[37] J. Li, C. Blake, D. S. J. De Couto, H. I. Lee, and R. Morris. Capacity of ad hoc wireless networks. In *7th Annual International Conference on Mobile Computing and Networking (MobiCom)*, 2001.

[38] K. Li, R. Kumpf, P. Horton, and T. Anderson. A quantitative analysis of disk drive power management in portable computers. In *USENIX Symposium*, 1994.

[39] L. Li, J. Y. Halpern, P. Bahl, Y.-M. Wang, and R. Wattenhofer. Analysis of a cone-based distributed topology control algorithm for wireless multi-hop networks. In *Principles of Distributed Computing (PODC)*, 2001.

[40] N. Li, J. C. Hou, and L. Sha. Design and analysis of an mst-based topology control algorithm. In *IEEE INFOCOM*, 2003.

[41] Q. Li, J. Aslam, and D. Rus. Online power-aware routing in wireless ad-hoc networks. In *7th Annual International Conference on Mobile Computing and Networking (MobiCom)*, 2001.

[42] J. R. Lorch and A. J. Smith. Reducing processor power consumption by improving processor time management in a single-user operating system. In *2nd Annual International Conference on Mobile Computing and Networking (MobiCom)*, 1996.

[43] J. P. Monks, V. Bharghavan, and W.-M. Hwu. A power controlled multiple access protocol for wireless packet networks. In *IEEE INFOCOM*, 2001.

[44] A. Muqattash and M. Krunz. Power controlled dual channel (PCDC) medium access protocol for wireless ad hoc networks. In *IEEE INFOCOM*, 2003.

[45] S. Narayanaswamy, V. Kawadia, R. S. Sreenivas, and P. R. Kumar. Power control in ad-hoc networks: Theory, architecture, algorithm and implementation of the COMPOW protocol. In *European Wireless 2002*, 2002.

[46] M. Pearlman, Z. J. Haas, P. Sholander, and S. S. Tabrizi. On the impact of alternate path routing for load balancing in mobile ad-hoc networks.

In *1st ACM International Symposium on Mobile Ad Hoc Networking and Computing (MobiHoc)*, 2000.

[47] C. E. Perkins and E. M. Royer. Ad-hoc on-demand distance vector routing. In *2nd IEEE Workshop on Mobile Computing Systems and Applications*, 1999.

[48] N. Poojary, S. V. Krishnamurthy, and S. Dao. Medium access control in a network of ad hoc mobile nodes with heterogeneous power capabilities. In *IEEE International Conference on Communications (ICC)*, 2001.

[49] J. Rabaey, J. Ammer, J. L. da Silva Jr., and D. Patel. PicoRadio: ad-hoc wireless networking of ubiquitous low-energy sensor/monitor nodes. In *IEEE Computer Society Annual Workshop on VLSI (WVLSI'00)*, 2000.

[50] R. Ramanathan and R. Rosales-Hain. Topology control of multihop wireless networks using transmit power adjustment. In *IEEE INFOCOM*, 2000.

[51] T. S. Rappaport. *Wireless Communications: Principles and Practice.* Prentice Hall, 1996.

[52] V. Rodoplu and T. H.-Y. Meng. Minimum energy wireless networks. *IEEE Journal on Selected Areas in Communications*, 17(8):1333–1344, August 1999.

[53] C. Schurgers, V. Tsiatsis, S. Ganeriwal, and M. Srivastava. Topology management for sensor networks: Exploiting latency and density. In *3rd ACM International Symposium on Mobile Ad Hoc Networking and Computing (MobiHoc)*, 2002.

[54] K. Scott and N. Bambos. Routing and channel assignment for low power transmission in pcs. In *5th IEEE International Conference on Universal Personal Communications*, 1996.

[55] R. C. Shah and J. M. Rabaey. Energy aware routing for low energy ad hoc sensor networks. In *IEEE Wireless Communications of Networking Conference (WCNC)*, 2002.

[56] E. Shih, P. Bahl, and M. J. Sinclair. Wake on wireless: An event driven energy saving strategy of battery operated devices. In *8th Annual International Conference on Mobile Computing and Networking (MobiCom)*, 2002.

[57] S. Singh and C. S. Raghavendra. Pamas: Power aware multi-access protocol with signalling for ad hoc networks. *ACM SIGCOMM Computer Communication Review*, 28:5–26, July 1998.

[58] S. Singh, M. Woo, and C. S. Raghavendra. Power-aware routing in mobile ad hoc networks. In *4th Annual International Conference on Mobile Computing and Networking (MobiCom)*, 1998.

[59] P. Sinha, R. Sivakumar, and V. Bharghavan. CEDAR: a core-extraction distributed ad hoc routing algorithm. In *IEEE INFOCOM*, 1999.

[60] I. Stojmenovic and X. Lin. Power-aware localized routing in wireless networks. *IEEE Transactions on Parallel and Distributed Systems*, 12(11):1122–1133, November 2001.

[61] M. W. Subbarao. Dynamic power-conscious routing for MANETs: An initial approach. *Journal of Research of the National Institute of Standards and Technology*, 104:587–593, November-December 1999.

[62] H. Tagaki and L. Kleinrock. Optimal transmission ranges for randomly distributed packet radio terminals. *IEEE Transactions on Communications*, 32(3):246–257, March 1984.

[63] C.-K. Toh. Associativity-based routing for ad hoc mobile networks. *Wireless Personal Communication*, 4(2):103–139, March 1997.

[64] C. K. Toh. Maximum battery life routing to support ubiquitous mobile computing in wireless ad hoc networks. *IEEE Communications Magazine*, 39(6):138–147, June 2001.

[65] Y.-C. Tseng, C.-S. Hsu, and T.-Y. Hsieh. Power-saving protocols for ieee 802.11-based multi-hop ad hoc networks. In *IEEE INFOCOM*, 2002.

[66] R. Wattenhofer, L. Li, P. Bahl, and Y.-M. Wang. Distributed topology control for wireless multihop ad-hoc networks. In *IEEE INFOCOM*, 2001.

[67] Y. Xu, S. Bien, Y. Mori, J. Heidemann, and D. Estrin. Topology control protocols to conserve energy in wireless ad hoc networks. Technical Report CENS Technical Report 0006, Center for Embedded Networked Sensing, January 2003. submitted to IEEE Transactions on Mobile Computing.

[68] Y. Xu, J. Heidemann, and D. Estrin. Geography-informed energy conservation for ad hoc routing. In *7th Annual International Conference on Mobile Computing and Networking (MobiCom)*, 2001.

[69] W. Ye, J. Heidemann, and D. Estrin. An energy-efficient mac protocol for wireless sensor networks. In *IEEE INFOCOM*, 2002.

[70] W. H. Yoen and C. W. Sung. On energy efficiency and network connectivity of mobile ad hoc networks. In *International Conference on Distributed Computing Systems (ICDCS)*, 2003.

[71] R. Zheng, J. C. Hou, and L. Sha. Asynchronous wakeup for ad hoc networks. In *4th ACM International Symposium on Mobile Ad Hoc Networking and Computing (MobiHoc)*, 2003.

[72] R. Zheng and R. Kravets. On-demand power management for ad hoc networks. In *IEEE INFOCOM*, 2003.

Chapter 7

USE OF SMART ANTENNAS IN AD HOC NETWORKS

Prashant Krishnamurthy
Dept. of Information Science and Telecommunications
University of Pittsburgh
prashant@tele.pitt.edu

Srikanth Krishnamurthy
Dept. of Computer Science and Engineering
University of California, Riverside
krish@cs.ucr.edu

Abstract The capacity of ad hoc networks can be severely limited due to interference constraints. One way of using improving the overall capacity of ad hoc networks is by the use of smart antennas. Smart antennas allow the energy to be transmitted or received in a particular direction as opposed to disseminating energy in all directions. This helps in achieving significant spatial re-use and thereby increasing the capacity of the network. However, the use of smart antennas presents significant challenges at the higher layers of the protocol stack. In particular, the medium access control and the routing layers will have to be modified and made *aware* of the presence of such antennas in order to exploit their use. In this chapter we examine the various challenges that arise when deploying such antennas in ad hoc networks and the solutions proposed thus far in order to overcome them. The current state of the art seems to suggest that the deployment of such antennas can have a tremondous impact in terms of increasing the capacity of ad hoc networks.

Keywords: Directional Antennas, Medium Access Control, Routing

7.1 Introduction

The use of smart antennas in cellular networks has been shown to offer an increased capacity by reducing interference and enabling spatial reuse of spectrum. Typically these antennas are deployed at base-stations in these networks to sectorize cells and focus transmissions in certain directions [1] [17]. A *smart antenna* usually consists of an array of antenna elements that work together in order to either focus the transmitted energy in a particular desired direction or to provide uncorrelated receptions of signals that can then be combined by complex signal processing techniques to improve the received signal quality or both. The spacing between the antenna elements is on the order of the wavelength of the carrier used for communications. Consequently, as technology makes the use of higher frequencies feasible, the spacing between the antenna elements can be much smaller. As an example, if the elements of the antenna array were to be arranged in a cylindrical layout, the radius of the cylindrical array would be just 3.3 centimeters if the ISM 5.8 Ghz band were used [16]. Similarly, if we were to use the 24 Ghz band, the radius of such a cylindrical array needs to be just around 0.8 centimeters. This in turn allows the use of small antenna elements that can be housed on mobile terminals. One could now potentially use these antennas in ad hoc networks that simply consist of mobile devices without a fixed supporting infrastructure.

The use of these antennas in mobile ad hoc networks however raises a new set of challenges. Traditional protocols (medium access control, routing and transport in particular) do not take advantage of the existence of the underlying antennas. Furthermore, in order to use the antennas effectively, support from the higher layer protcols is necessary. There has been a lot of recent interest in the design of new protocols to facilitate the use of smart antennas in ad hoc networks.

In this chapter, we review the current state of the art protocols at the medium access control and the routing layers and discuss why they are appropriate for use with smart antennas in ad hoc networks. We elaborate on the problems that they are capable of solving, discuss their limitations, and identify problems that are yet to be completely solved. We begin with a brief discussion of smart antennas and models that are typically used in studies thus far. We then discuss solutions at the medium access control layer that have been proposed for use with such antennas. Finally, we investigate the challenges that arise at the routing layer and the work to date on this topic.

7.2 Smart Antenna Basics and Models

In this section, we describe the different kinds of antenna systems and discuss their characteristics in brief. It is not our intent here to describe the signal processing techniques required for tuning antenna patterns or the assoiciated subject matter in electromagnetics and the interested reader is referred to [11] and [12]. As mentioned earlier, our goal in this chapter is to primarily look at the networking challenges that arise due to the deployment of such antennas in ad hoc networks.

7.2.1 Antennas in Brief

Omni-directional antennas are those antennas that radiate or receive energy equally well in all directions. Traditionally these antennas have been considered (or implicitly assumed) for studies related to ad hoc networking. Since these antennas dissipate energy in all directions, they impose limitations on the extent to which the wireless spectrum may be re-used in the network. In [15], it was shown that the capacity of an ad hoc network that uses omni-directional antennas is limited.

Smart antennas, naively speaking, have the ability to receive/transmit energy in a particular direction as compared to other directions. The energy dissipated in the directions other than the desired direction can be quelled when transmitting and filtered out while receiving. Smart antennas also null out the interference caused by other transmissions. The antenna is complemented by an adaptive array processor that decides on the amount of power to be used on each antenna element so that the signals combine together to form a specific antenna pattern. The lack of such signal processing techniques causes energy to be dissipated in directions other than the desired one.

An antenna that simply *beamforms* the energy in a particular direction is often referred to as a *directional antenna*. Most of the work to date has looked primarily at the use of directional antennas in ad hoc networks. Figure 7.1 depicts the antenna patterns of (a) an omni-directional antenna and (b) a directional antenna. The antenna footprint of a directional antenna contains a *main lobe* and *side lobes* as shown in the figure. The Yagi antenna [13] is a well known directional antenna often used in cellular networks. It has been shown that the capacity of ad hoc networks can be increased significantly by using directional antennas [18].

We further classify directional antennas systems into *switched beam* (or sectorized) antenna systems and *steerable beam* systems. In switched beam systems only multiple fixed beams are possible. As an example, space might be divided into four sectors of 45° each. A directional transmission would then cover one of these four fixed sectors. A given node in the network cannot focus its antenna beam on a particular neighborhood node so as to maximize the signal strength at that node. Clearly, for a switched beam antenna with K beams, the width of each beam is $2\pi/K$ radians. In contrast, in a steered beam system, the main lobe of the antenna can be focused in practically any desired direction. Thus, if a given node is communicating with its neighbor, it can adaptively steer its beam so as to point the main lobe towards that neighbor in a mobile scenario as well.

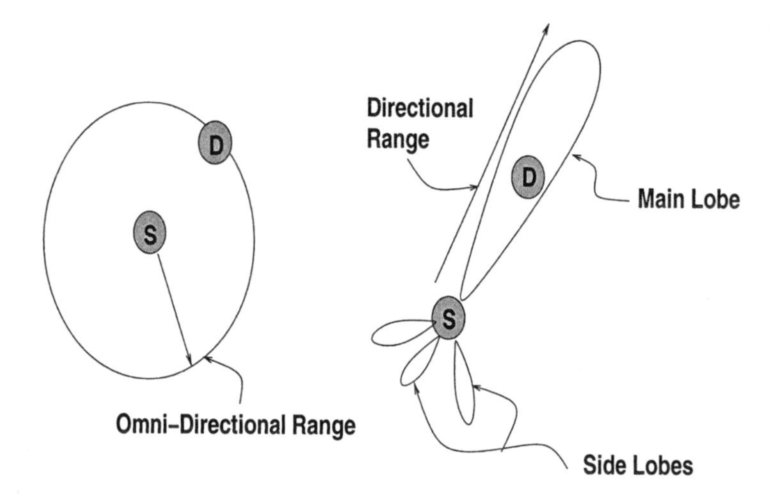

Figure 7.1. Footprint of (a) An Omni-directional Antenna and (b) A Directional Antenna

The nomenclature *smart antennas* typically refers to more sophisticated antenna arrays. There are dynamic phased arrays that maximize the gain towards a target in the presence of multi-path effects and there are adaptive arrays that can produce nulls so as to eliminate the effects of simultaneously ongoing interfering transmissions.

We point out that the antennas that we consider for deployment in ad hoc networks are *electronically steerable* antennas. High-gain aperture and horn antennas that are commonly used in satellite or microwave based terrestrial wireless networks are inappropriate for use in mobile terminals. These antennas will have to be mechanically steered and could be extremely expensive in terms of the energy consumed; this in turn could significantly increase the energy usage in the mobile terminal and could quickly cause its battery to die.

7.2.2 Important Antenna Parameters

The gain of a directional antenna is typically higher than that of an omni-directional antenna. Correspondingly directional antennas can have a higher reachability or in other words, a larger *directional* range as compared to an omni-directional antenna. The gain of a directional antenna is defined as [16]:

$$G_d = \frac{\eta U(d)}{U_{ave}} \tag{2.1}$$

where, η is defined to be the efficiency of the antenna and accounts for the hardware related losses, $U(d)$ is the energy in the direction d and U_ave is the average power density. The gain is measured in dBi, where i is used to indicate that this is the gain in decibels over an ideal isotropic antenna.

The main lobe of the antenna represents the direction of *peak gain* during a transmission or a reception. The antenna beamwidth typically corresponds to the angle subtended by the two directions on either side of the peak gain that are 3 dB lower in gain as compared to the peak gain. Note that this is a reduction by half in terms of the signal power as compared to the power in the direction of peak gain (not in decibels). This angle is also sometimes referred to as the *3 dB beamwidth.*

7.2.3 Directional Antenna Models

The presence of side-lobes causes interference to other simultaneous transmissions in spite of using a directional antenna. Sophisticated antenna arrays can steer these side lobes so as to create nulls towards other simultaneous users of the channel. However, simpler (and hence cheaper) antennas suffer from the presence of these sidelobes. In ad hoc network literature, most of the work thus far adopts one of two models for characterizing the radiation pattern of a directional antenna: (a) The Flat Topped Radiation Pattern and (b) The Cone and Sphere Radiation Pattern.

With the flat topped radiation pattern model, it is assumed that the gain of the antenna is a constant within a defined beamwidth of radiation. It is also assumed that the side lobes are absent. If the beamwidth is θ, the gain is computed to be $G = 2\Pi/\theta$.

With the cone and sphere radiation pattern, the side lobes are accounted for by a spherical footprint that is attached to the apex of a cone. The axis of this cone passes through the direction of peak gain of the antenna. If the gain in the direction of the main lobe and the beamwidth of the main lobe are known, it is a simple exercise to compute the gain of the spherical side-lobe in the cone and sphere radiation pattern [16]. We depict the cone and sphere radiation pattern in Figure 7.2.

7.3 Medium Access Control with Directional Antennas

In this section, we describe the current state of the art literature on medium access control with directional antennas. Medium access control refers to the arbitration of channel bandwidth among a plurality of multiple-access users. We can classify medium access protocols into two types (a) on-demand or unscheduled access and (b) scheduled access. On-demand or unscheduled access mechanisms are based on *contention* access. Nodes in the ad hoc network contend for the channel. Carrier Sensing (both virtual and physical) are used to reduce the extent of packet losses due to collisions. Traditionally, the MAC protocol defined in the IEEE 802.11 standard has been popularly adopted as the

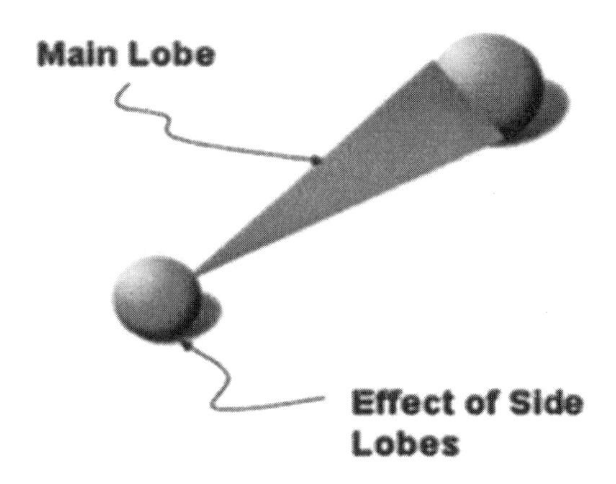

Figure 7.2. The Cone and Sphere Radiation Pattern

contention-based MAC protocol in ad hoc network research. Scheduled access, on the other hand, attempts to schedule transmissions in advance to reduce the possibility of collisions. Protocols that use scheduled access might proactively allocate bandwidth based on a number of criteria that may include the topology, the generated traffic and priority of various nodes.

We begin with a very brief discussion of the IEEE 802.11 MAC protocol and then go on to point out the problems that one would face if this protocol is used "as is" with directional antennas. We then discuss various approaches that have been proposed for addressing these problems. Finally, we discuss the few approaches that have been proposed for scheduled access.

7.3.1 The IEEE 802.11 MAC Protocol in Brief

The Distributed Co-ordination Function (DCF) specified in the IEEE 802.11 MAC standard has been popularly advocated for ad hoc networks. The DCF function is based on co-ordinating medium occupancy using carrier sense multiple access with collision avoidance (CSMA/CA). The approach alleviates the hidden terminal problem that arises in wireless networks by the use of a simple CSMA scheme.

In the IEEE 802.11 MAC protocol, a transmitter sends a Request to Send (RTS) message to a recipient neighbor when it wishes to send a data packet to that neighbor. The RTS message implicitly informs the neighboring nodes within the omni-directional range of the transmitting node that a data transfer is being initiaited. If possible, the receiver would then respond with a Clear to Send or CTS message. The CTS message implicitly informs the nodes in the neighborhood of the receiver of the forthcoming data transfer. The RTS-CTS handshake is then followed with the transmission of the data (DATA) and the acknowledgement (ACK) messages.

All four frames (RTS,CTS,DATA and ACK) contain information about the duration of the communication. Neighbors that overhear any of these messages back off from performing any transmissions during the specified period. This is ensured by what is called *vrtual carrier sensing*. Each node maintains a *Network Allocation Vector* (NAV) which contains information about the state of the channel in the vicinty of the node. When a node overhears a handshake, it retrieves the duration specified in the packet and updates its NAV so as to preclude transmissions until the communication indicated by the handshake is completed. Thus, a node is permitted to transmit only if its NAV is equal to zero. If the NAV is a positive number, there is a countdown until it reaches zero. Note that subsequent handshakes can increase the NAV. The virtual carrier sensing is used in conjunction with physical carrier sensing in order to reduce the possibility of collisions.

7.3.2 Directional Transmissions and the IEEE 802.11 MAC protocol

With directional transmissions and receptions, it is now possible for nodes to send and receive data in specific directions. Clearly, for maximum spatial re-use it is desirable that all communications be directional. However, this may prevent some nodes from knowing the existence of an on-going communication. This could potentially lead to collisions. On the other hand, omni-directional transmissions or receptions of messages may limit the spatial re-use possible in spite of using directional antennas. Various combinations of directional and omni-directional transmissions and receptions have been considered and the trade-offs that arise from the use of such combinations have been studied [10] [14], [16], [2], [6], [21]. We discuss the variants in this sub-section.

What do the RTS and CTS messages mean now ?. The RTS and CTS messages may be either transmitted directionally or omni-directionally. However, the receipt of an omni-directionally transmitted control message by an overhearing node no longer implies that the particular node ought to be prohibited from performing transmissions. On the other hand, if we resort to directional transmission of control messages, the non-receipt of a control message no longer means that a node can initiate transmissions [16]. In order to elucidate this, we describe the examples considered in [16]. Two scenarios are shown in Figure 7.3.

In the first scenario, it is assumed that the RTS message is transmitted omni-directionally. When A transmits its RTS message to B, the message is overheard

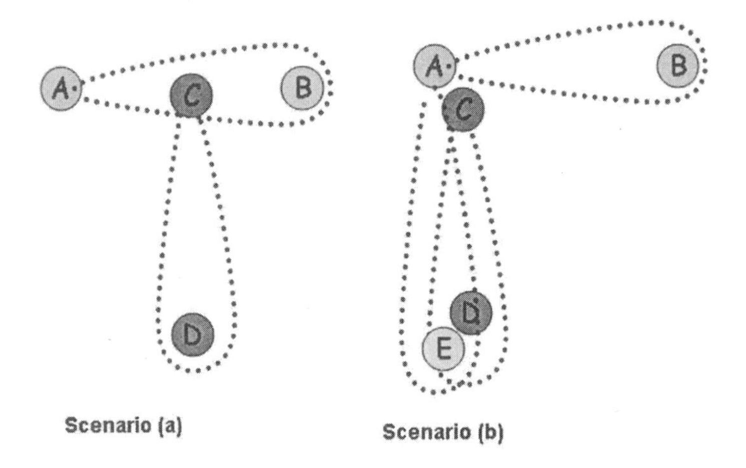

Figure 7.3. The effect of omni-directional / directional transmissions of control messages with the 802.11 MAC Protocol

by C. C then updates its NAV and defers its transmission to D until the communication between A and B is complete. However, clearly, C could have initiated its communication to D without interfering with the communication between A and B. This shows that the receipt of the RTS message by C did not necessarily imply that C should not indulge in transmissions.

In the second scenario, we assume that the RTS messages are transmitted directionally. Thus, A does not hear the RTS message sent out by node C while it is in communication with node B. In the meantime, C has begun the transmission of its data packet to node D. Once node A completes its communication with node B, it initiates a *new* communication with node E. This causes a collision at node D. Note that in this case, despite the fact that node A did not hear an RTS message, its new handshake caused a collision.

7.3.3 Directional Medium Access Control with Omni-Directional Receptions

In [16], Ramanathan considers two approaches to deal with this problem. The first approach which is called the *conservative approach* precludes a node from performing transmissions upon the receipt of any control message. The second approach which they call the *aggressive approach*, allows a node to initiate new transmissions in spite of hearing control messages sent by other nodes. RTS and CTS messages are assumed to be transmitted and received omni-directionally. The RTS and CTS messages are assumed to contain location information of both the sender and receiver; this in turn helps transmit (or receive) the DATA and ACK messages directionally. Neither of the two schemes overcome the problems discussed above. The performance evaluation of the two schemes in [16] shows that both schemes outperform the IEEE 802.11 MAC with omni-directional communications. The aggresive scheme was found to be better since the conservative scheme suffered from extremely high latencies due to the nodes deferring their communications over extended periods even when the channel was free in reality.

In [10], Ko *et al* also study various modes of transmitting the RTS and CTS messages. The receptions are all considered to be omni-directional. They classify their protocols as Directional MAC or D-MAC protocols. The authors assume that the transmitter is aware of the location[1] of the receiver node. They consider two schemes; the first scheme is based on the use of a directional RTS message while the second scheme allows the use of both directional and omni-directional transmission of RTS messages. The CTS messages are transmitted omni-directionally. The range extension possible due to the use of directional antennas has not been considered.

In Scheme 1, the sender node transmits a directional RTS (DRTS) message to the recipient neighbor. If this message is successfully received, in response, the receiver transmits an omni-directional CTS (OCTS) message. The OCTS message contains the location of the node sending the OCTS message as well as the location of the node that initially sent out the DRTS message. After the successful handshake, the sender node sends out the DATA message directionally and in response, the receiver sends an ACK message back to the sender, also directionally. We refer the reader to Figure 7.4 for the following discussion. The example is based on [10] and helps illustrate the functionalities of D-MAC.

Consider a communication between nodes B and C. Node B sends a DRTS that is received by Node C. Node A does not receive the message and hence is free to transmit. Node C's OCTS message is heard by Node D. Node D however is aware of the location of Node C (thanks to the information in the OCTS message) and precludes transmissions only in the direction of C. Nodes that are within the range of the DRTS message from B are precluded from transmissions in the direction of B. They are free to communicate with nodes in other directions. However, it is important to note that a node might respond to a DRTS message only if none of its antenna directions are *blocked* since with this scheme, the node is required to send the CTS omni-directionally.

In Scheme 2, omni-directional transmission of RTS messages are considered in addition to the transmission of DRTS messages. The use of DRTS messages as in Scheme 1 could sometimes cause collisions. To illustrate this, we once again consider the scenario in Figure 7.4. Since A is unable to receive the DRTS message sent out by Node B, it could potentially initiate a new transmission

[1]Note that this might be possible using the global positioning system (GPS). However, in inhospitable terrain this may not be feasible. We discuss possible neighbor discovery and maintenance methods later

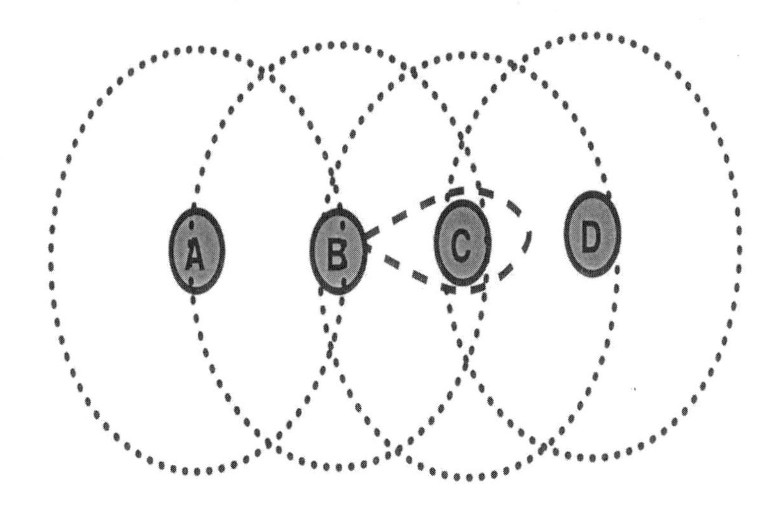

Figure 7.4. A Scenario to Understand the Schemes Proposed in D-MAC

to Node B (using a DRTS message). This transmission might collide with the OCTS or the ACK message sent out by Node C. In Scheme 2, in order to alleviate this effect, omni-directional transmissions of RTS messages are permitted. When a particular node wishes to initiate communications, it first checks to see if any of its directional antennas are blocked. If none of the antennas are blocked, it sends out an omni-directional RTS message. If not, and if the antenna that points in the desired direction is unblocked, the node sends out a DRTS message using that antenna. If the desired antenna is blocked, then the node would defer its transmission until the antenna was free.

The performance of the proposed schemes are evaluated in [10] using simulations. It is shown that both Scheme 1 and Scheme 2 outperform the traditional IEEE 802.11 MAC protocol using omni-directional transmissions. It is seen that at light loads, Scheme 2 outperforms Scheme 1 since it reduces the possibility of collisions. However, at heavier loads, the conservative nature of Scheme 2 reduces the efficiency of spectrum re-use (although some robustness to collisions is still provided) due to the omni-directional transmission of RTS messages. Thus, the throughput achieved is lower with Scheme 2 than with Scheme 1.

7.3.4 Adding directional receptions: Directional Virtual Carrier Sensing

Takai *et al* proposed the use of direectional virtual carrier sensing (DVCS) in [14]. In the previously discussed schemes, the receipt of an RTS or a CTS message is simply used as a criterion that dictates the deferral of transmissions for the duration indicated in the message. The reason that this was deemed appropriate was that the nodes simply performed omni-directional receptions. However, if directional receptions were possible, one could potentially examine *directional channel availability*. Thus, a node, upon the receipt of an RTS or a CTS message, precludes transmissions only in those directions in which it interferes with the communications related to the received RTS or CTS message. The mechanisms proposed in [14], like in other work discussed earlier, are extensions to the IEEE 802.11 MAC protocol to enable its use in the presence of directional antennas. In the following paragraphs, we discuss these mechanisms in brief.

First, for directional virtual carrier sensing to work, each node would need to cache estimated angles of arrival (AOAs) from neighboring nodes whenever it hears any signal from these nodes. This is done regardless of whether the communication was intended for the node under discussion. The AOA is an indicator of where the node is located. Thus, when a node wishes to communicate with a particular neighbor, it uses the cached AOA information with regards to that neighbor and sends a directional RTS message in the the direction of the neighbor. Additional attempts are made if the node fails to receive a CTS response from the neighbor. If the node does not get a CTS response after 4 directional RTS attempts, it resorts to omni-directional transmissions of the RTS message. The cached AOA information is purged after the failure of the directional RTS attempts. In order to comply with the IEEE 802.11 standard where the total number of RTS attempts for a particular packet are limited to seven, the node will make three omni-directional attempts before it drops the packet and reports a link failure to the higher layers.

The authors in [14] assume that a steerable beam antenna is used. Upon the receipt of an RTS message, the receiver is assumed to *lock* the receive beam pattern for maximizing the received power. Similarly, the orignal transmitter does the same upon the receipt of the CTS message from the receiver. The beam patterns are then used both for transmission and reception. Upon the completion of the communication (through the transmission of the ACK message) the beam patterns are *unlocked*. The locking prevents the nodes from listening to other transmissions for the duration of the communication.

Finally, each node that overhears a control message exchange, uses a directional network allocation vector (DNAV) as opposed to the network allocation vector (NAV) used in the original IEEE 802.11 MAC protocol. The DNAV allows a node to specify each direction (the direction of the main lobe of the communication as determined from the control message) and the angular beamwidth in each particular direction that is to be avoided so as to remove conflicts with ongoing communications. When a node wishes to initiate a new transmission (or respond to a new request for communications) it checks its DNAV to see if the particular requested direction is open and transmits in the direction only if it is. The width of DNAV is dictated by the beamwidth scoped by the underlying directional antenna. If this beamwidth can be dynamically changed, the DNAV would take this into account.

The performance studies in [14] show that the use of DVCS can provide up to a threefold or fourfold increase in network capacity (measured in terms of throughput) as compared to the IEEE 802.11 protocol used with omni-directional antennas. They also do additional simulation experiments to study the behavior of the scheme in the presence of some nodes that are simply capable of omni-directional transmissions and they find that there is still a benefit in using DVCS (only nodes equipped with directional antennas use DVCS).

7.3.5 The impact of increased directional range

The work in [14] assumes that the directional range is equal to the omni-directional range. However, as mentioned earlier, in reality, the directional range might be much higher than the omni-directional range. This is actually beneficial since network partitions that may actually occur with the use of omni-directional antennas could be potentially bridged by the use of directional antennas. Furthermore, since the range is longer, one can now potentially compute shorter routes which can lead to improved efficiency.

However, the medium access control protocol discussed earlier has certain problems when an extended directional range is considered. In [6], Roy Choudhury, Yang, Ramanathan and Vaidya look at these problems in depth. They call the basic medium access control protocol the Directional MAC or DMAC protocol; the protocol is similar to the protocol described in [14]. The protocol assumes that nodes are aware of the locations of their neighbors. When idle they listen in the omni-directional mode. However, when a node intends to transmit a message it transmits a directional RTS message towards the intended neighbor. A directional CTS is sent in response. Both directional transmissions and directional receptions are employed for the exchange of the DATA and the ACK messages. Overhearing nodes update their directional network allocation vectors as in the protocol in [14].

The basic DMAC suffers from a number of problems due to the directional nature of the communications and the increased directional range. Note here that if the gain of the omni-directional antenna is G_o and the gain of the directional antenna for a given beamwidth is G_d, then $G_d > G_o$. Furthermore, if one were to deploy directional transmissions but omni-directional receptions, the total gain seen is $G_d G_o$. On the other hand, if one were to deploy both directional transmissions and directional receptions the gain observed would be G_d^2.

In order to understand the problems, we consider the scenario shown in Figure 7.5. The scenario is similar to the one considered in [6]. First we

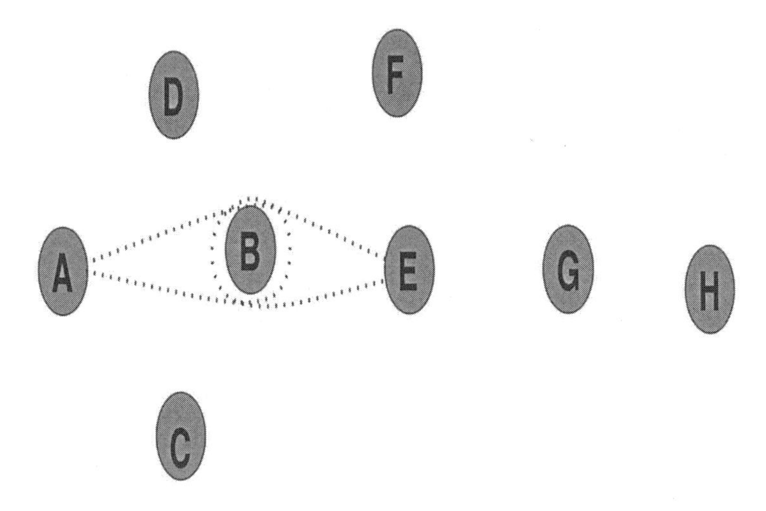

Figure 7.5. A Scenario to Understand the Problems with DMAC

consider the problem of *hidden terminals due to unheard RTS/CTS messages.* In the scenario, let B initiate a transmission to D. Subsequently, E might initiate a data transmission to G or vice versa. Note that even though B might be in the directional range of node G, it does not receive this CTS message. Thus, upon the completion of its communication with D, B might attempt a transmission to node G thereby causing a collision at E. Note that carrier sensing does not help here since B cannot physically sense the communications between E and G. Thus, when involved in directional communications, a node might miss out on hearing some of the RTS or CTS messages. Upon the completion of its communication it might initiate new transmissions that would interfere with the communications related to the missed RTS/CTS messages.

The second problem that we consider is the problem of *hidden terminals due to asymmetry in gain.* In order to discuss this problem, we once again refer to the example in Figure 7.5. We consider an example wherein node B iniates a communication with node E. The handshake is achieved by the exchange of a DRTS and a DCTS message (from B and C respectively). If A is in the omni-directional reception mode, it is possible that it does not hear the DCTS message sent by node E. Note that the total antenna gain in this case is $G_d G_o$. Once the data communication between nodes B and F begins, let us assume that node A wishes to initiate a communication with node B (clearly it is unaware of the communication already in progress). Node A now sends an RTS directionally in the direction of node B. Node E's antenna is beamformed to receive in the direction of A. The antenna gain between nodes A and E is now G_d^2 since both the transmission and the reception are directional. Thus it is possible that node A's signal now reaches node E and this would cause a collision at node E (between the data from B and the DRTS from A).

The third problem that was identified in [6] was the problem of *deafness*. We once again refer to Figure 7.5. Consider the case wherein node D is sending data to node E via node B. The directional exchange of control messages might not be heard by node C. During the time that B is transmitting the message from D to E, node C might attempt to transmit a DRTS message to node B. However, since node B has beamformed in the direction of E, it is unable to receive the RTS. Hence, C does not receive a CTS response. In accordance to the IEEE 802.11 MAC protcol policy, node C would then back off. If node D were to have a continuous stream of packets destined for node E, this problem might repeat itself. Node C would continue to experience RTS failures and would increase its back-off interval. This phenomenon, referred to as deafness, could therefore cause *false link failures* (C believes that the link to B has failed even if it has not) and unfairness in channel access.

Finally, due to the higher gain of directional antennas, the shape of the regions where transmissions are blocked (referred to as *silenced regions* in [6]) are different for omni-directional and directional communications. When both are used, the silenced regions vary depending upon the traffic and the network topology. The authors of [6] do not examine this in detail in the paper. Quantifying the trade-offs while using hybrid directional/omni-directional communications has still not been explored in detail.

7.3.6 The Multi-hop RTS MAC Protocol (MMAC)

Roy Choudhury *et al* attempt to to exploit the increased directional range via the Multi-hop RTS MAC protocol (MMAC) in [6]. The basic problems with hidden terminals and deafness still exist with the MMAC protocol. However, the authors claim that the benefits due to the exploitation of the increased range somewhat compensates for the other negative effects. To recap, if both the sender and the receiver are beamforming (i.e, both directional transmissions and directional receptions are invoked) the antenna gain can be potentially much higher than in the case where they use directional transmissions but omni-directional receptions or vice versa. In Figure 7.6 if all the nodes were listening omni-directionally, node A would be able to communicate (with a directional transmission) with only nodes D and B. However, if node E were to be receiving directionally, node A could communicate with node E.

The basic idea in MMAC is to *route* an RTS message via multiple hops to the intended recipient asking the recipient to beamform in the direction of the originator of the RTS message. The neighbors of a node are divided into two types: (a) The Direction-Omni (DO) Neighbors are those neighbors of a node that can receive transmissions from the node even if they are in the omni-directional

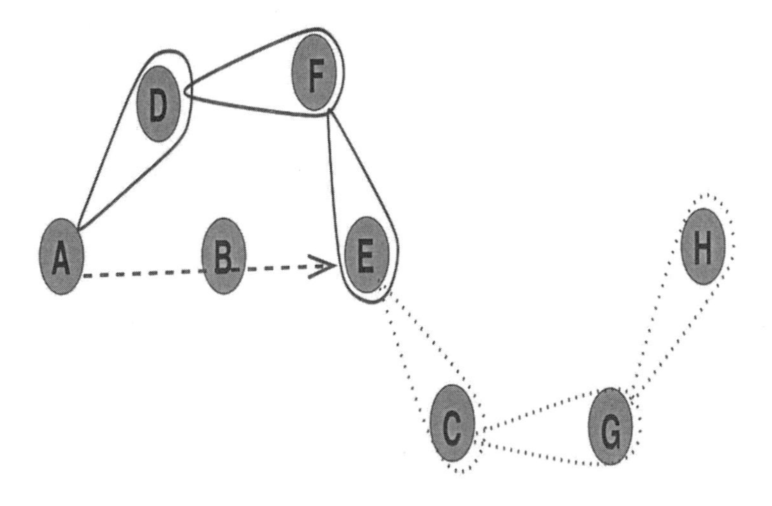

Figure 7.6. The MMAC Protocol

reception mode. (b) The Direction-Direction (DD) neighbors of a node are those neighbors that can hear from the node only if they are beamformed in the direction of the node. Thus, a DD neighbor of a node (say node A) cannot hear from node A if it is receiving information in the omni-directional mode. On the same note, one can also think of (i) an Omni-Omni range (OOR) wherein a transmission and the reception are both omni-directional, (ii) a Direction-Omni range (DOR) where the transmission is directional but the reception is omni-directional and (iii) a Direction-Direction range (DDR) where both the transmission and reception are directional. Typically the OOR is the smallest and the DDR is the largest. The idea behind MMAC is to form links between DD neighbors. The advantage of doing this is to reduce the hop-counts on routes and in bridging possible network partitions.

A DD neighbor of a node may be also be reached via multiple-hops through other neighbors of the node. Typically, the nodes on such a route are DO neighbors of each other and such a route is referred to as the DO-neighbor route. This DO-neighbor route is used to request the DD-neighbor of interest (the receiver) to point its receive beam in the direction of the DRTS transmitter at a future time.

We describe the MMAC with the help of an example; towards this we refer the reader to Figure 7.6 In this example, node A is the initiating transmitter. The objective is to send a message to node H. If each node were to use its DO neighbors to forward the packet, the route from A to H could be potentially 6 hops. However, if the DD neighbors were to be used, the path could be shortened to two hops (A to E and E to H). In order to communicate directly with its DD neighbor E, node A uses the DO route to E. In [6], the authors assume that a higher layer at node A is aware of the DO-neighbor route[2]. The route in this case would be specified to be via nodes D and F.

In order to ensure that the channel is reserved for its communication with E, A would first send out an RTS message in the direction of E. The duration field in this RTS message takes into account the entire duration of the communication including the multi-hop RTS transmissions and the following CTS, DATA and ACK transmissions. This time for the multi-hop transmission of RTS messsages is calculated as the product of the time required for a single RTS transmission and the number of hops on the multi-hop route. The RTS message specifies the destination to be E. A node that overhears this RTS message (for example, node B in this case) would set its DNAV in the direction of A and in the opposite direction of E. Thus, if D^{AB} specifies the direction towards A, B also sets its DNAV in the direction specified by $(D^{AB} + 180) mod 360$ (in degrees). If the destination of the RTS, viz. E, happens to receive the DRTS message from A directly (it is possible that it is beamformed in the direction of A), it would switch to the omni-mode to be able to receive the multi-hop RTS. Alternatively, it could simply send back a CTS to A right away but this was not considered in [6].

Node A then would send a special type of RTS message which is called the *forwarding* RTS message and forwards it on to D, which in turn relays it to F and so on. The forwarding RTS message contains the entire DO-neighbor route to node E. Note that in order to transmit this forwarding RTS message the same rules that govern the basic DMAC are to be followed (i.e., the physical carrier sensing and the directional virtual carrier sensing should both indicate that the channel is free for transmission). If a node receives or overhears the forwarding RTS message it does not alter its DNAV. Each node on the route gives the highest priority for the transmission of the forwarding RTS message (i.e., unlike in the IEEE 802.11 specification, the nodes do not back-off upon sensing the channel to be free). If a DO-neighbor is busy or has the DNAV set in the direction in which the forwarding RTS is to be transmitted, it simply drops the RTS. Note also that the forwarding RTS message is not responded to by a CTS message or acknowledged in any other way.

Meanwhile, node A (after completing its forwarding RTS transmission to node D) beamforms in the direction of node E and awaits a CTS. If no CTS is received, it times-out and initiates the whole process again. The time-out is caclulated on the basis of the time needed for the forwarding RTS message to traverse the DO-neighbor route and for the recipient (node E) to respond with a CTS. If node E receives the multi-hop RTS correctly, it responds with a CTS in the direction of node A. The transmission of the CTS is preceded by both

[2]The practicality of MMAC hinges on this assumption. Protocols that have been proposed so far for performing routing with directional antennas will be discussed later

physical and virtual carrier sensing as in DMAC. After the CTS is received by A, it proceeds to send the DATA packet directionally and this is followed by a directional ACK from node E to node A. Nodes that overhear either the CTS or the DATA messages update their DNAVs accordingly.

The authors in [6] perform extensive simulations to study the performance of DMAC and MMAC. They find that in topologies where nodes are aligned (either string topologies wherein nodes are arranged along a line or in grid topologies) the benefits of using the directional antennas are dwindled due to the problems of deafness and asymmetry described earlier. The benefits were more pronounced when random topologies were considered. One of the limitations of this work was that the authors assume that a node is aware of its neighborhood and somehow has the routing information required to send out the multi-hop RTS messages. Furthermore, the protocols are vulnerable to deafness and do not study neighbor discovery and the tracking of neighbors in mobile scenarios.

7.3.7 Dealing with Deafness: The Circular RTS message

In [21], the authors propose the use of *the circular RTS message* to deal with many of the problems reported in [6]. Omni-directional transmission and reception of the RTS messages could result in directional neighbors not knowing about the forthcoming communication since the OOR is potentially much smaller than than DOR. However, simply using a directional RTS could potentially result in the hidden terminal and deafness problems reported in [6]. Korakis, Jakllari and Tassiulas propose that instead of transmitting the directional RTS in simply the direction of the intended neighbor the RTS be now transmitted in all possible directions. To illustrate this we refer to Figure 7.7.

Note in the figure that by *circularly* transmitting the RTS message in each of the M possible directions, a node can potentially inform all of its DO neighbors of its intended transmission. The source also indicates the antenna beam (switched beam antennas are assumed) on which the intended transmission is to take place. Accordingly, nodes can (a) set their DNAV vectors appropriately (b) recognize that the node is in the process of communication and avoid the problems due to deafness. Each node is required to maintain a location table where it records the information with regards to the communications in progress and the directions in which these communications are being carried out.

When transmitting the circular RTS message a node has to take care not to transmit the message in those directions where it is prohibited from doing so (due to either physical or virtual carrier sensing). Thus, the circular RTS message cannot completely eliminate the problems due to hidden terminals and deafness. The authors perform extensive simulations to show that in spite

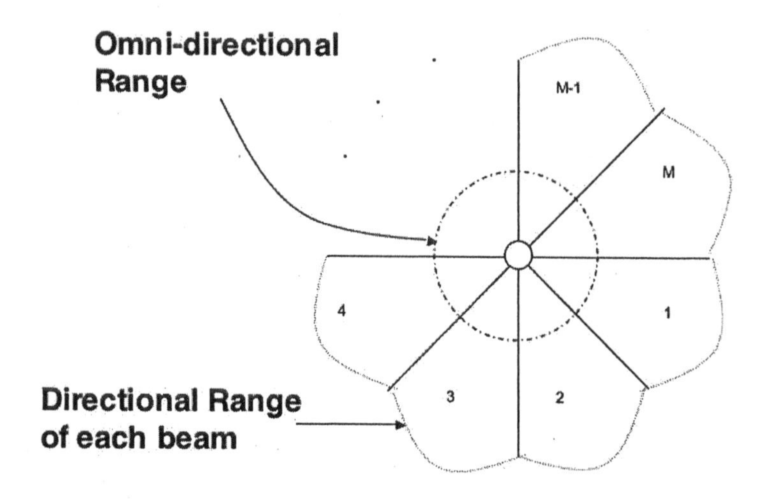

Figure 7.7. The Circular RTS message

of this, in typical scenarios, the circular RTS message helps alleviate these problems to a large extent. The unfairness in access seen with DMAC is also reduced to a large extent.

The circular RTS is also extremely useful in tracking neighbors in mobile conditions. Since the node transmits the RTS message in all possible directions, even if a neighbor has moved, it can still possibly hear the RTS message and respond with a CTS message. Thus, the new proposed medium access control scheme is robust under mobility to a large extent.

One of pitfalls of using the circular RTS message is that there is an additional latency incurred with every transmission. If a neighbor were to successfully transmit an RTS message in all of the M possible directions (Figure 7.7), the time required is M times that required for a single RTS transmission. Furthermore, this scheme generates a significant amount of overhead by transmitting these multiple directional RTS messages. In spite of these limitations the use of the circular RTS is the only proposed scheme to date that reduces the effects of hidden terminals and deafness with directional antennas.

7.3.8 Other Collision Avoidance MAC Protocols

There are other MAC protocols designed for use with directional antennas [3], [19], [20]. The protocols are similar to the ones described. The key ideas are based on nodes identifying the directions in which there are ongoing communications and supressing transmissions in those directions until the present communications are completed. While [3] suggests marking the antenna sector on which the transmission was received (sectorized antennas are assumed) for achieving this, [20] suggests the use of explicit information in the control messages to indicate the direction of transmission. We do not discuss these schemes further.

7.3.9 Scheduled Medium Access Control

The MAC protocols described thus far are based on collision avoidance. These protocols suffer from high collision rates when the load is high. An alternative approach is to have scheduled access wherein nodes exchange control messages that allow them to know of each other's traffic patterns and thereby somehow schedule collision-free (to the extent feasible) transmissions. There has been little work on scheduled access with directional antennas and we describe the work to date from [13] by Lichun Bao and J.J.Garcia-Luna-Aceves. In this paper, a new protocol called the Receiver Oriented Multiple Access (ROMA) has been proposed for scheduled access with directional antennas. One other difference in this work as compared with other efforts is that the authors assume the presence of *multi-beam antenna arrays* (MBAA). The advancement of digital signal processing technologies facilitate the use of such arrays. With an MBAA a node can generate multiple beams that allow the node to communicate with more than one of its neighbors. It is assumed that the MBAA can generate up to K transmit antenna beams. The radiation pattern of an MBAA may be depicted as shown in Figure 7.8. The MBAA also has the ability to anull radiations in unwanted directions.

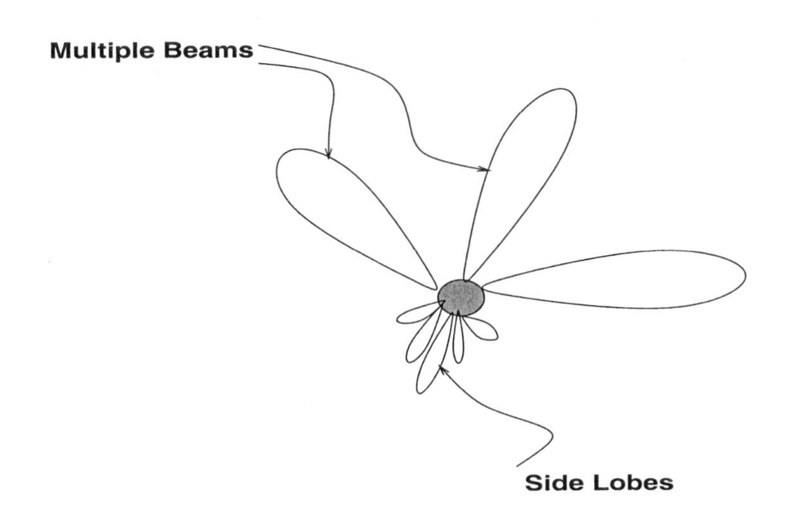

Figure 7.8. The Multi-Beam Antenna Array

The authors assume that the MBAA system is capable of transmitting to multiple neighbors but is capable of making just a single reception at any given time. Furthermore, they assume that the system is capable of performing omni-directional transmissions and receptions. They consider a time-slotted system i.e., time is divided into contiguous frames. The nodes are assumed to have a synchronized view of time by using either the global positioning system (GPS) or the network time protocol (NTP). Each node is assumed to know the precise location of its one-hop neighbors.

Each node then propagates its one hop neighbor information to all of its one-hop neighbors. Thus, this propagation gives each node knowledge of its *two-hop* neighborhood. In order to propagate this information, the authors assume that the nodes use omni-directional random access transmissions. Receptions are omni-directional as well. In order to accommodate this, the authors split time into segments. In the scheduled access segment the time is further divided into slots and access in these slots is in accordance to a schedule to be described later. In the random access segment, nodes exchange the control information. In [13] the authors do a simple analysis to compute the fraction of time needed for the random access and show that this is fairly small.

The scheduled access takes the following scenarios into account: (a) avoidance of hidden terminal problems wherein a recipient node ends up receiving transmissions from two simultaneous senders that are hidden from each other (b) ensures that the schedule respects the half-duplex nature of the communications (c) two transmitters are not trying to reach the same receiver at the same time. Each node then depending on its own identifier (ID) and a time-slot identifier computes a priority for itself. This priority is based on the use of a simple hash function. Similarly it computes priorities for each of its neighbors. Depending on the traffic generated and its relative priority, a node will make a decision on whether or not to transmit in a particular scheduled slot. Note that the aforementioned scenarios are to be taken into account while this decision is being made. A similar computation is made on the links on which a node would transmit. As a simple example, if a node is of lower priority, it might be unable to transmit on a subset L of its K possible links since there are higher priority nodes using those links.

In addition to this priority assignment, a node will also have to either take the role of a transmitter or a receiver during each slot. If the calculated priority is even, then the node decides to be a receiver and if it is odd, it chooses to transmit. There could be pathological cases wherein a node and all of its neighbors are all either transmitters or receivers. In such a case, the node from the group that has the highest priority will switch its configuration; in other words, if there is a particular group created such that a node and all of its neighbors are receivers, the node with the highest priority in that group will switch to being a receiver.

ROMA offers collison free access and has been shown to perform well. However, mobile scenarios are not considered. Furthermore, the priorities based on which the schedules are formed are based on identifiers and not based on the traffic generated.

7.4 Routing with Directional Antennas

The use of directional antennas can have an effect on routing. On-demand routing schemes can now scope their route queries in the direction in which the destination was last seen. With omni-directional antennas multi-path routing wherein (multiple paths are found between a source and a destination and used simultaneously) cannot be exploited very well since packets routed on one of the paths cause an interference zone that typically encompasses the other paths and thereby limits the number of packets routed on these paths. With directional antennas it is now possible to construct disjoint paths that do not interfere with each other [6]. The scheduling of transmissions (the directions in which antennas are to be pointed at different times) is tightly coupled with routing. However, current state of the art research has not looked at routing in great depth. It still remains an open area of research and possibilities for joint MAC/routing layer optimizations remain. In this section, we review the work on routing to date.

7.4.1 On Demand Routing Using Directional Antennas

The first work on routing with directional antennas was by Nasipuri et al [2]. In this work, the authors examine the impact of directional antennas on the performance of on-demand routing protocols (such as the Ad hoc On Demand Distance Vector Routing or the Dynamic Source Routing [9]). On-demand routing protocols are based on *searching* for a route to a desired destination when the need arises. This search typically involves the flood of a route request or RREQ message. The key idea in [2] is to propagate this route request message in the *direction* of the desired destination with the help of directional transmissions by a restricted set of nodes. The authors assume the presence of simple switch beam or sectorized antennas. Two protocols are proposed.

In the first protocol, when a source (say S) intends to compute a new route to a destination denoted by D, it broadcasts the route request query in the direction in which it had been communicating earlier with D. Any node that receives this query would then use the same technique, i.e., propagates the query in the same direction. This in effect, causes the query to be flooded in a conical section in the presumed direction of the destination. Clearly, the advantage of this process is to limit the scope of the flood. The scheme has been designed with the premise that the destination would not have moved too far from its initial position when it communicated with the originating node S. If this query were to fail, the query is re-initiated. The second time, it is flood throughout the network. The main drawback of this protocol is that it requires that the destination be in the same directional sector as the first hop on the path. If there exist circuitious routes wherein the destination is in a direction that is different from that of the preliminary search regime, then the preliminary search would fail.

In the second proposed protocol in [2], the authors propose that when a particular route is found, the source should record the directions of the antennas used at each hop on the route. The relays that return the response from the route query from the destination add this information to the response packet header. This allows a node to get a rough estimate of the direction in which the destination is located depending upon the hop-count on the path and the number of times a particular direction was used. If a particular direction was used more than others, then the authors suggest that the particular direction be used in order to initiate the directional query. Clearly, the proposed schemes can lead to unsuccessful directional query floods. However, the authors show by simulations that the advantages in terms of the reduction in the quantum of overhead via successful directional floods outweigh the wasteful overhead due to unsuccessful floods.

7.4.2 The Impact of Directional Range on Routing

The increased range of directional antennas can actually help in terms of reducing the number of hops needed in order to reach a destination i.e., can help in establishing shorter routes. Furthermore, in scenarios where omni-directional transmissions may result in partitioned disjoint subnetworks, the extended range can help in bridging the sub-networks. In [5], Roy-Choudhury and Vaidya examine the impact of directional antennas on routing. The authors first perform simulations to understand the impact of directional antennas on routing. Based on their observations, they propose strategies that can exploit the presence of directional antennas and further analyze their new strategies via simulations. They assume that the DMAC protocol described earlier (proposed in [6]) is used in conjunction with the routing protocols. They then use the Dynamic Source Routing (DSR) protocol [8] over the DMAC protocol to study its performance.

The DSR protocol is an on-demand routing protocol proposed for ad hoc networks. The protocol was designed with the premise that omni-directional transmissions and receptions are employed. We provide a very brief overview of DSR. A source broadcasts a route query message in order to find a destination. Nodes that hear the query broadcast it further; if they have a cached route to the destination they respond with a response instead of furthering the query. The destination upon receiving a query sends a response back to the source with a choice of the path. The identities of the relays on the entire route is recorded in the response packet. When a node wishes to use the route for sending data, it records the entire route in the packet header (hence the name source routing). When a route fails, the node that discovers the failure sends a route error message back to the source. This stimulates the re-initiation of a route query at the source.

To recap, a node had both DO neighbors and DD neighbors. The DO neighbors were those neighbors that could be reached via directional transmissions but omni-directional receptions and the DD neighbors were those that could be reached only if directional receptions were being used in addition to the directional transmissions. The OO neighbors were defined to be those neighbors that can communicate via omni-directional communications. Omni-directional broadcast of control messages may not result in the discovery of the *shortest* routes since only the OO neighbors would be reached. Thus, in order to reach a DO neighbor or a DD neighbor, the broadcast will have to be relayed via an OO neighbor. This in turn would result in the discovery of paths that are potentially longer than those that are possible i.e., a path that is much shorter thanks to the extended reach of directional communications might never be found. As an example, in Figure 7.9, if one were to only use omni-directional transmissions of route requests, the route to C from A would always be via B (a two hop route). If on the other hand, one could somehow use directional requests, the direct link from A to C could be found. Furthermore, if the destination node belongs to a separate network partition that can only be reached via directional communications omni-directional transmission of control messages would fail to discover the destination.

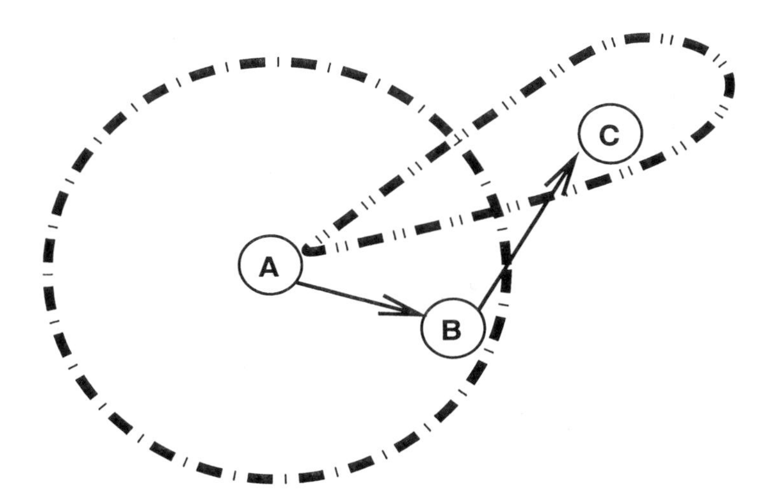

Figure 7.9. Impact of omni-directional route requests

In order to ensure that shorter routes via DO neighbors are reached, the authors in [5] propose the concept of *sweeping*. The idea is to transmit the route request directionally in all possible directions. This is akin to the circular

RTS message transmissions described in the previous section. This helps in transmitting the route query directly to the DO neighbors as opposed to via OO neighbors. The authors also propose a scheme similar to the circular RTS message to cope with mobility. They propose the transmission of HELLO messages directionally on each antenna beam. This process is referred to as *scanning*. When a node receives the HELLO message from a neighbor it responds to the message using the appropriate antenna beam. Scanning can be expensive and in order to restrict the scope of scanning the authors propose to use what is called *partial scanning*. If a neighbor moves out of its directional range, the HELLO messages are now sent out only on the K beams that are adjacent to the beam that was previously in use for that neighbor. K is a system parameter that can be set based on the conditions of mobility.

In their simulation studies the authors in [5] find that if the distance between the source and the destination was small, then there was not much to be gained due to the increased range (as one might expect). If the source and the destination were further apart, then the gains due to the increased directional range were more evident. With increased densities, the gains were not significant either; this has been attributed due to the increased interference effects at these densities due to the presence of side lobes.

It was also found that the route request messages can experience excessive delays due to sweeping. Note that the duration of a sweep in N directions is equivalent to the duration of N separate sequential transmissions. Furthermore, while sweeping a node starts with a random direction. Consequently, it is possible for the route query that traverses the best path to arrive at the destination later than a route query that traverses a longer sub-optimal path. In order to overcome this effect, the authors propose what is called the *delayed route reply optimization*. Upon receiving the first route request query, the destination would wait a pre-specified time (a system parameter based on the time it takes for a complete sweep) before it responded to the route request query. It collects all the route request queries that are received within this time and chooses the best route recorded from among the records in the received queries. It then sends back a response to the route with this best route.

The authors observe that the overhead incurred in terms of performing sweeping is also excessive and much higher than that incurred with DSR. In order to reduce the overhead, the authors propose that the route request be forwarded in directions opposite to the direction in which the original request was received. As an example, in Figure 7.10, Node A receives the RTS on Beam 1 and forwards it on the beams opposite to Beam 1 viz., beams 2 3 and 4 respectively. Similarly, node B will forward the request only in the range of directions shown. This process is called the *selective forwarding optimization* process and is found to reduce the overhead incurred due to sweeping significantly.

However, the overhead still remains higher than in DSR using omni-directional communications.

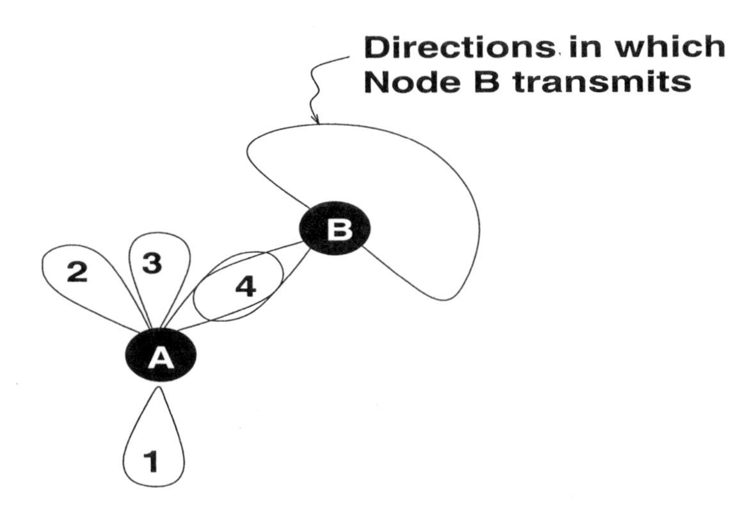

Figure 7.10. The Selective Forwarding Optimization

Finally, in [5], the authors find that the problems with deafness (a direct consequence of DMAC) remain and the performance of the routing schemes in terms of throughput seem to be poor with regular topologies. With random topologies however, they observe significant gains in throughput in spite of the increase in overhead due to sweeping.

7.4.3 A Joint MAC/Routing Approach

In [6], S.Roy et al design a new routing protocol that attempts to compute multiple paths and balance the load across the multiple paths. Directional antennas are assumed. Once the multiple paths are found, it becomes important to choose the right path for a connection since one can get the maximum out of the network if the interference zones created by the transmissions on the path taken a connections were disjoint to the extent possible with the interference zones created by other connections. In order to illustrate this point we refer Figure 7.11. We have two sources S_1 and S_2 and these nodes want to establish connections with nodes D_1 and D_2, respectively. For the connection from node S_1 to node D_1 two distinct routes are feasible. The first is via nodes N_1 and N_2 and the second via nodes N_5 and N_6. If the former route is chosen it creates high levels of interference to the second connection that is being routed via nodes N_3 and N_4. This problem of two paths that can create severe levels of interference to each other is called *route coupling*. Link state information (in the form of lists) are exchanged in order to facilitate an awareness of the routing activities in the neighborhood. The nodes then choose the paths that are *zone disjoint* from the paths are already in use. For shorter paths (less than a pre-specified hop-count) the zone-disjointness is ignored and the shortest path is simply chosen. However, a more careful assessment is made for longer paths. The policy is thus chosen since the increase in path length when computing zone disjoint paths for nodes that are close to each other is significant and may in fact increase the levels of interference experienced by connections that start later.

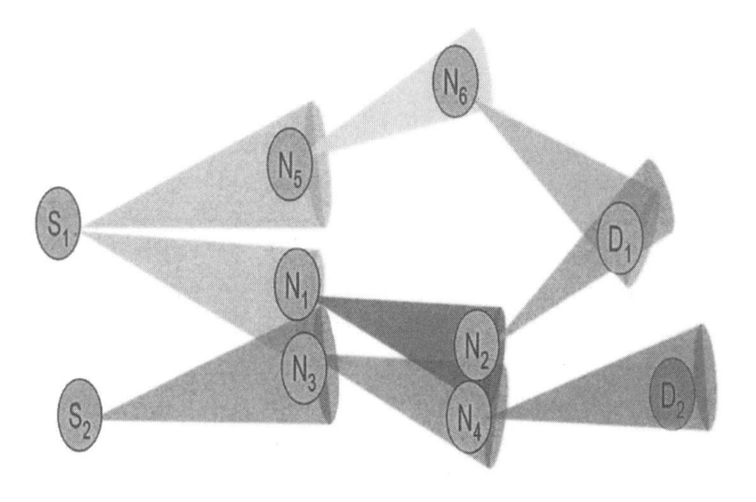

Figure 7.11. Route Coupling

7.4.4 Remarks

In order to efficiently use directional or smart antennas a unified MAC/Routing approach is needed. The exchange of information between how transmissions are scheduled and how routes are chosen are tightly coupled. Methods that can overcome problems with regards to tracking mobile terminals and that can overcome deafness are needed. As pointed out in [5] the use of directional antennas can provide significant benefits; however, in some scenarios, if proper care is not taken, the use of these antennas can in fact cause a degradation in performance.

7.5 Broadcast with Directional Antennas

In mobile ad hoc networks, it is often required to send a broadcast packet to all nodes in the network. This is called *network-wide broadcasting* or simply *broadcasting* in the literature. For example, several route discovery protocols assume that there is a method by which packets can be propagated with information about the route from one node to all other nodes in the network. It is common to assume that such a mechanism exists without actually considering its impact. In a simple scenario, a node will broadcast a packet to all neighbors. In turn they will broadcast it to their neighbors and so on. If a broadcast packet has already been received and transmitted, it is not broadcast again by a node if it receives a duplicate copy. If many neighboring nodes receive a broadcast packet from a source node at approximately the same time, they may try to rebroadcast it simultaneously resulting in collisions. To avoid this problem, nodes will wait for a random amount of time before they rebroadcast the packet. An additional possibility is to track redundant receptions of the same broadcast packet during this random delay. If the node is aware of the topology, it need not rebroadcast the packet if it has already received it from all of its neighbors. Several enhancements to this simple scheme have been proposed. They have been compared in [4]. In this section we will not look at these broadcast schemes, but we will describe problems with broadcasting and how recent research has tried to use directional antennas to address these problems. The assumption here is that the IEEE 802.11 MAC protocol is used. We do not consider scheduling based or centralized MAC protocols here.

7.5.1 Performance Issues in Broadcasting

We can broadly classify performance issues in evaluating any broadcasting scheme as follows [4]- [7].

1 *Redundancy:* If simple flooding is employed, it is likely that several transmissions of the broadcast packet will be redundant, especially when nodes that rebroadcast are close together.

2 *Contention and Collisions:* Even with random delays, it is possible that there will be increased contention for transmitting broadcast packets resulting in collisions. Hidden terminals can exacerbate the problem because the RTS/CTS mechanism is not employed when a packet needs to be broadcast. A node will simply wait for the channel to be clear before transmitting the broadcast packet.

3 *Coverage:* As a result of contention and collisions, some nodes may never receive the broadcast packet if the lifetime is exceeded.

4 *Latency:* This is the time taken for the broadcast packet to reach all the modes in the network.

5 *Impact of traffic types:* In a network where both broadcast and unicast traffic are present, the load of one may affect the load of the other.

6 *Algorithm efficiency:* Any scheme that attempts to reduce the problems arising from broadcast traffic must be efficient to implement. For in-

stance, given the network topology, it is possible to use brute force to determine a minimum connected set of rebroadcasting nodes that could broadcast the packets to all nodes not in the set. The problem of determining this minimum set is NP hard and not suitable for practical implementation.

7 *Impact of mobility:* A broadcasting scheme may perform very well in a static network but it may degrade when nodes are moving. It is important to consider the effects of mobility.

8 *Impact of network topology:* The performance of broadcasting schemes can vary drastically depending on how dense the network is and what the topology may be. The performance in a network where nodes are arranged in a rectangular grid may be very different from the performance when nodes are randomly scattered in a region.

Considering the performance issues enumerated above, the research question of interest is whether directional antennas can be exploited in a broadcast scheme that reduces redundancy, latency, contention and collisons, improves coverage, is robust under mobility and is simple to implement. The research literature on broadcast schemes in ad hoc networks making use of directional antennas is fairly sparse. We discuss two different research papers below.

7.5.2 Broadcast schemes with directional antennas

In [22], simulations of two directional broadcast schemes are performed. The ad hoc network consists of nodes arranged on concentric circles. Each node is assumed to be capable of using a directional antenna of a given beamwidth. Nodes are half duplex and cannot receive and transmit at the same time. A node transmits CBR traffic with unicast and broadcast packets such that broadcast traffic forms some percentage r of the total number of packets. The performance measure that is used is the throughput of unicast and broadcast traffic by considering only nodes in a local area thereby reducing boundary effects. In this work, three different 802.11-like MAC protocols are simulated. In the first case, both RTS and CTS packets are transmitted using omnidirectional antennas. In the second case, the RTS is transmitted in a given direction only whereas the CTS packet is transmitted in all directions. In the last case, both RTS and CTS packets are transmitted in given directions only. The description of the protocol is not clear in some respects. It is not clear whether all packets (unicast and broadcast) use directional antennas or whether only the control packets employ directional antennas. The directionality of receiving antennas is not specified. It is also not clear whether the gain of directional antennas is considered in the simulations. The simulations do indicate that as the node density increases, using directional RTS and directional CTS provides the best throughput for both unicast and broadcast traffic. The effect is more perceptible when the beamwidth is smaller. Otherwise, all three schemes statistically have the same performance. There is a huge variation in the observed throughputs making it important to consider the network topology.

Three different broadcast schemes with increasing complexity are proposed in [7] and simulated using Qualnet. The benchmark for comparison is the simple scheme when all nodes use omnidirectional transmissions. With omnidirectional antennas, nodes blindly forward a broadcast packet to all neighbors if the packet has not been already received previously and the maximum hop count has not been reached. In the case of broadcast schemes making use of directional antennas, it is assumed that the nodes use 90° *switched beam* antennas with the same gain as the omnidirectional antenna so that any bias in the broadcast performance due to the extended range of the directional antennas is minimized. It is assumed that all nodes can detect the angle-of-arrival of packets. In all of the schemes, no control RTS/CTS packets are used. The simulations also do not take into account the simultaneous presence of unicast traffic in the network. A random mobility model is used to study the impact of mobility as well.

The first scheme is called *on/off directional broadcast.* Here, a node that receives a broadcast packet will rebroadcast it only in directions other than the one from which the packet arrived. If the same packet arrives from more than one direction before the random delay expires (because of a rebroadcast from another node), all of these directions are ignored in the rebroadcast. In the second scheme called *relay node directional broadcast*, a node that broadcasts a packet will pick exactly one node per direction as a relay node. The broadcast packet will let the recipient know if it is a relay node and also indicates the situation when no relay node is selected. the latter could happen when the transmitting node is not aware of a neighboring node in that direction. The relay node is usually te farthest node in a given direction and it is determined using periodic hello packets. This scheme also makes use of the techniques of the on/off scheme by switching off broadcasts in the directions from which broadcast packets have been already received. Finally, a *location based directional broadcast* scheme is proposed where nodes are aware of their positions by being equipped with a GPS or some other positioning system. A transmitting node will inform other nodes of its location in the broadcast packet. this information is used by a relay node (as in the second scheme) to change the random delay of transmission in its beams. The beam that is expected to cover the largest *extra area* has the smallest delay. This way, a rebroadcasting node waits longer in directions that may have received the broadcast packet already.

Simulations in [7] over a 1500 $\times$ 1500 m area with 100 nodes indicated that using directional antennas in a static environment improves coverage and reduces redundancy and collisons regardless of the scheme compared to the omnidirectional scheme. The location based directional broadcast however has a larger latency due to the fact that the rebroadcast is performed in different directions sequentially with the random delays being much larger than with the other schemes. the advantage of the location based directional broadcast is that it provides the best coverage. The relay node directional broadcast scheme has the smallest latency and the on/off directional broadcast has the smallest number of collisions and redundant packets. When the network becomes mobile, the relay node directional broadcast has poorer coverage than the omni-directional scheme because of the need to determine a relay node in a given direction. The conclusions in this work are that on/off directional broadcast scheme has low complexity and reasonably improved performance. It appears to be a suitable candidate for directional broadcast.

7.6 Summary

In this chapter we review the current literature on protocols that are designed for use with smart antennas in ad hoc networks. Most of the protocols designed thus far assume that the antennas are simple directional antennas. The work on medium access control involves using the antennas intelligently for collision avoidance and spatial re-use purposes. Most of the protocols are based on *directional virtual carrier sensing* wherein overhearing nodes simply preclude transmissions in the direction of an ongoing communication as opposed to not transmitting in any direction as would be the case if a omni-directional antenna were to be used. The impact of directional antennas on routing was discussed. The increased transmission range due to directional communications has been shown to be beneficial in terms of computing shorter paths and bridging partitions. To conclude, although there has been some work on protocols that exploit the use of directional antennas in ad hoc networks, the protocols cannot completely overcome some of the problems that arise due to the use of such antennas and are not yet designed to aptly cope with mobility. Challenges remain and the area remains an exciting and open for future research.

References

[1] A.Acampora and S.V.Krishnamurthy. A new adaptive mac layer protocol for broadband packet wireless networks in harsh fading and interference environments. *IEEE/ACM Transactions on Networking*, 2000.

[2] A.Nasipuri, J.Mandava, H.Manchala, and R.E.Hiromoto. On-demand routing using directional antennas in mobile ad hoc networks. *IEEE IC-CCN*, 2000.

[3] A.Nasipuri, S.Ye, J.You, and R.E.Hiromoto. A MAC protocol for mobile ad hoc networks using directional antennas. *IEEE WCNC*, 2000.

[4] B.Williams and T.Camp. Comparison of broadcasting techniques for mobile ad hoc networks. *ACM MOBIHOC*, 2002.

[5] R.Roy Choudhury and N.H.Vaidya. Impact of directional antennas on ad hoc routing. *IFIP Personal and Wireless Communications (PWC)*, 2003.

[6] R.Roy Choudhury, X.Yang, R.Ramanathan, and N.H.Vaidya. Using directional antennas for medium access control in ad hoc networks. *ACM MOBICOM*, 2002.

[7] C.Hu, Y.Hong, and J.Hou. On mitigating the broadcast storm problem in MANET with directional antennas. *IEEE ICC*, 2003.

[8] D.Johnson, D.Maltz, and J.Broch. Dynamic source routing for mobile ad hoc networks. *IETF MANET Working Group*, 1998.

[9] E.Royer and C-K.Toh. A review of current routing protocols for ad-hoc wireless networks. *IEEE Personal Communications Magazine*, 1999.

[10] Y.Ko *et al.* Medium access control protocols using directional antennas in ad hoc networks. *IEEE INFOCOM*, 2000.

[11] J.C.Liberti and T.S.Rappaport. Prentice Hall, 1999.

[12] J.Kraus and R.J.Marhelka. McGraw Hill, 2002.

[13] L.Bao and J.J.Garcia-Luna-Aceves. Transmission scheduling in ad hoc networks with directional antennas. *ACM MOBICOM*, 2002.

[14] M.Takai, J.Martin, A.Ren, and R.Bagrodia. Directional virtual carrier sensing for directional antennas in mobile ad hoc networks. *ACM MOBIHOC*, 2002.

[15] P.Gupta and P.R.Kumar. The capacity of wireless networks. *IEEE Transactions on Information Theory*, 1998.

[16] R.Ramanathan. On the performance of ad hoc networks with beamforming antennas. *Proceedings of ACM MOBIHOC*, 2001.

[17] S.V.Krishnamurthy, A.Acampora, and M.Zorzi. Polling based media access protocols for use with smart adaptive array antennas. *IEEE//ACM Transactions on Networking*, 2001.

[18] S.Yi, Y.Pei, and S.Kalyanaraman. On the capacity improvement of ad hoc wireless networks using directional antennas. *ACM MOBIHOC*, 2003.

[19] T.ElBatt and B.Ryu. On the channel reservation schemes for ad-hoc networks utilizing directional antennas. *IEEE International Symposium on Wireless Personal Multimedia Communications*, 2002.

[20] T.ElBatt, T.Anderson, and B.Ryu. Performance evaluation of multiple access protocols for ad hoc networks using directional antennas. *IEEE WCNC*, 2003.

[21] T.Korakis, G.Jakllari, and L.Tassiulas. A mac protocol for full exploitation of directional antennas in ad hoc wireless networks. *ACM MOBIHOC*, 2003.

[22] Y.Wang and J.J.Garcia-Luna-Aceves. Broadcast traffic in ad hoc networks with directional antennas. *IEEE GLOBECOM*, 2003.

Chapter 8

QOS ISSUES IN AD-HOC NETWORKS

Prasun Sinha
Computer and Information Science
Ohio State University
Columbus, OH 43210
prasun@cis.ohio-state.edu

Abstract Support for QoS is integral to the design of ad-hoc networks. Fluctuations in channel quality effect the QoS metrics on each link and the whole end-to-end route. In addition, the interference from non-neighboring nodes effects the link *quality*. QoS is thus an essential component of ad-hoc networks. The QoS requirements arise at the application layer in the form of restrictions on values of certain QoS metrics. The most commonly studied QoS metrics are bandwidth, delay and jitter. Bandwidth is the QoS metric that has received the most attention in the QoS literature. The QoS requirements are typically met by soft assurances rather than hard guarantees from the network. Most mechanisms are designed for providing relative assurances rather than absolute assurances. This chapter presents solutions and approaches for supporting QoS in ad-hoc networks at the physical, MAC, and routing layers. It also presents approaches at other layers and describes future challenges that need to be addressed to design a QoS enabled ad-hoc network.

Keywords: QoS, Ad-hoc, 802.11e

8.1 Introduction

The need for supporting QoS in the Internet is evidenced by an increasing activity in the IETF community for supporting the Diffserv [3] architecture. The initial designers of the Internet moved away from the telephone network design where the intelligence was in the network and the end-terminals were comparatively dumb. The telephone network design however provides QoS in the form of guaranteed connection and quality of voice, once a call is established. The initial Internet design idea of keeping the network simple and moving the intelligence to the edge and the end-hosts, did help in the rapid growth of the Internet. But with proliferation of applications requiring some notion of guarantee of service from the network, it is becoming essential to support QoS in the Internet. Multimedia communication and VoIP (Voice over IP) are two such applications that are rapidly gaining popularity. The performance of these applications largely depends on the QoS assurances provided by the network.

In ad-hoc networks, Quality of Service support is becoming an inherent necessity rather than an "additional feature" of the network. Following are the three main reasons that make a strong case for designing QoS enabled ad-hoc networks rather than adding such features as an afterthought.

- **Wireless channel fluctuates rapidly and the fluctuations severely effect multi-hop flows.** As opposed to the wired Internet, the capacity of the wireless channel fluctuates rapidly due to various physical layer phenomena including fading and multi-path interference. In addition, background noise and interference from nearby nodes further effect the channel quality. In ad-hoc networks, the end-to-end quality of a connection may vary rapidly as change in channel quality on *any* link may effect the end-to-end QoS metrics of multi-hop paths.

- **Packets contend for the shared media on adjacent links of a flow.** Contention between packets of the same stream at different nodes impacts the QoS metrics of a connection. Such contention arises as the wireless channel is *shared* by nodes in the vicinity. Unlike in the Internet, this phenomenon effects the QoS even in the absence of any other flow in the network.

- **Interference can effect transmissions at nodes beyond the neighbors.** Interference effects are pronounced in ad-hoc networks where typically a single frequency[1] is used for communication in the shared channel. In single-hop infrastructured wireless networks frequency planning is mostly used where nearby base-stations can be configured to function at different frequencies for reducing interference. Transmissions in the wireless media are not received correctly beyond the transmission range. But even beyond the transmission range, the remaining power may be enough to interfere with other transmissions. So, interference from non-neighboring nodes may result in packet drops.

In order to support QoS on multi-hop paths, QoS must be designed for the end-to-end path as well as for each hop. The physical and MAC layers are responsible for QoS properties on a single-hop. The routing layer is responsible for QoS metrics on an end-to-end route.

[1] In some recent studies such as [17], the use of multiple frequencies has been explored for ad-hoc networks.

The concept of ad-hoc networking is not tied to any particular single-hop wireless technology. However, with increasing deployments of Wireless LAN (WLAN) devices at homes, offices and public hotspots, the term *wireless* is becoming synonymous with "Wireless LANs". Currently in the market there are products conforming to two competing WLAN standards, namely IEEE 802.11a and IEEE 802.11b. These standards differ from the original IEEE 802.11 standard in the specification of the physical layer. However, the MAC layer is unchanged in all these three protocols. In this chapter, we use 802.11 to collectively refer to the three standards. High speed (up to 54 Mbps with 802.11a), decreasing prices (Wireless Network Interface Cards are priced below $50) and proliferation of wireless integrated handheld devices, are the three main reasons for its popularity.

Most researchers assume CSMA/CA based 802.11 (specifies Medium Access (MAC) and Physical layers) to be the underlying wireless technology for ad-hoc networks. In this chapter we will also assume that 802.11 is the underlying technology. Researchers are also actively exploring the use of other medium access techniques such as TDMA [19], for ad-hoc networks. More recently, there has been a growing interest in applying ad-hoc networking techniques to different environments, such as acoustic ad-hoc networks [21] for marine exploration. Figure 8.1 shows a Wireless LAN and Figure 8.2 an ad-hoc network. The 802.11 standard has two modes of operation, namely the Infrastructure mode and the ad-hoc mode. These modes correspond to the WLAN and ad-hoc configurations respectively. In the WLAN configuration nodes communicate only via the access-point (AP). In the ad-hoc configuration, nodes communicate via multi-hop peer-to-peer wireless links formed by virtue of proximity with other nodes.

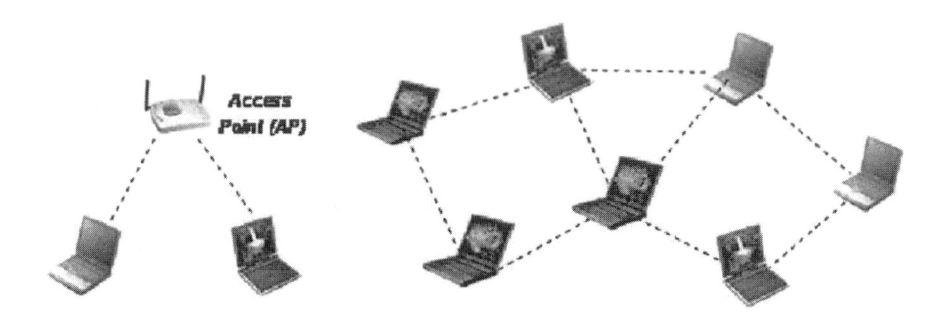

Figure 8.1. Wireless LAN

Figure 8.2. Ad-hoc Network

The rest of the chapter is organized as follows. Section 8.2 provides a definition of QoS and a discussion on QoS metrics. Section 8.3 presents QoS issues in the design of the physical layer. Section 8.4 discusses QoS support at the MAC layer in WLANs and ad-hoc networks. Section 8.5 describes various solutions for QoS routing in ad-hoc networks. Section 8.6 discusses other QoS approaches at transport and higher layers. Frameworks that span more than one networking layers are discussed in Section 8.7. Section 9 presents some future challenges in the design of QoS enabled Ad-hoc Networks and concludes the chapter.

8.2 Definition of QoS

So far we have been discussing about QoS in an abstract sense. But, what is QoS? What are the QoS metrics?

Quality of service refers to different notions at different networking layers. At the physical layer, QoS refers to the data rate and packet loss rate on wireless links, which is a function of the channel quality. With continuously varying channel quality, it is impossible to maintain constant data rate and low packet loss rate. At the MAC layer, QoS is related to the fraction of time a node is able to successfully access and transmit a packet. At the routing layer, end-to-end QoS metrics would depend on the metrics at each hop of a multi-hop route. The routing layer must try to compute and maintain routes that satisfy the QoS requirement for the lifetime of a connection. The transport and upper layers could include support for QoS if the routing layer is not able to meet the QoS requirements.

Bandwidth, delay and jitter are the three commonly studied QoS metrics. However, the problem of QoS in ad-hoc networks is more challenging than in wired networks as described in Section 8.1. As a result there has been little work on supporting delay and jitter; and most of the focus has been on providing bandwidth assurances. Various mechanisms have been proposed to estimate the amount of bandwidth in CSMA/CA (Carrier Sense Multiple Access) networks [8] and TDMA networks [12].

For ad-hoc networks, it is difficult to provide hard QoS guarantees due to fluctuations in the wireless channel and interference from non-neighboring nodes. It is therefore easier to design solutions where QoS support from the network is in the form of soft-assurances [18] rather than hard guarantees. For the same reasons, relative assurances are more common than absolute assurances. Most of this chapter refers to soft-assurances for QoS metrics, unless stated otherwise.

8.3 Physical Layer

One of the fundamental challenges in wireless networks is the continuously changing physical layer properties of the channel. The physical layers of 802.11a and 802.11b can support multiple data rates. Depending on the channel quality the data rate can be altered to keep the bit error rate acceptable, as high data rates are also prone to high bit error rates.

The 802.11a standard operates in the 5.7 GHz band and supports data rates of 6, 9, 12, 18, 24, 36, 48 and 54 Mbps, The 802.11b standard operates in the 2.4 GHz band and supports 1, 2, 5.5 and 11 Mbps. However, the standards do not specify any mechanisms to discover the highest possible rate on a link.

The data rate switching policy has a direct impact on the QoS metrics of the channel. For example, the most conservative switching policy of always staying at the lowest channel rate will guarantee equal physical layer data rate on all links in ad-hoc networks. If an application requires all links to have the same data rate, a policy of using the lowest data rate may work. However, this leads to severe under-utilization of resources as the links with good channel quality do not send at the highest possible rates.

For efficient use of a multi-rate physical layer, there has been several algorithms proposed at the physical layer. Some of these algorithms are closely tied to the MAC layer as well. They impact the observed throughput on a link and the end-to-end throughput of a multi-hop connection. The QoS requirements of upper layers may effect the design of this algorithm. However, the current proposals are all based only on improving the link utilization, although they may be modified to implement QoS requirements of higher layers.

8.3.1 Auto Rate Fallback (ARF)

[9] presents an algorithm called Auto Rate Fallback (ARF) for finding the highest possible data-rate on a wireless link. It was designed for Lucent's Wavelan II devices based on the IEEE 802.11b standard. The default operation is at the highest data-rate. When a MAC layer ACK is missed after successful transmissions, the first retransmission is done at the same rate. If the ACK is missed again, the rate is lowered to the next data-rate for subsequent transmissions and re-transmissions. If ten ACKs are received correctly or if a timer expires, then the device attempts to upgrade the data-rate. If the first transmission at the higher data rate fails, it immediately drops to the lower data-rate.

8.3.2 Receiver-Based Auto Rate (RBAR)

[7] observed that the data rate of a 802.11 link can fluctuate very frequently (on the order of 50 times per second) and the ARF algorithm is not capable of altering the data-rate according to the changing channel conditions. They propose a rate adaptive MAC protocol called RBAR (Receiver-Based Auto Rate). The algorithm makes use of the RTS-CTS exchange in 802.11 DCF mode to learn about the current condition of the channel. The SNR (Signal to Noise Ratio) of the RTS is used to determine the highest possible data-rate that can be used for DATA packets. The maximum allowed data rate is informed to the sender using the CTS. Since the channel estimation is done at the receiver just before the data transmission, the data-rate estimation is very accurate.

8.3.3 Opportunistic Auto Rate (OAR)

[15] proposes a mechanism called Opportunistic Auto Rate (OAR) for improving throughput in the presence of multi-rate links in ad-hoc networks. The key idea is to send multiple packets when the channel rate is higher. The RBAR protocol can be used to compute the channel rate that can be supported. Similarly, OAR can also be used with sender based rate adaptation protocols such as ARF. However, it has been shown that RBAR outperforms ARF [7]. The algorithm ensures that all nodes are granted channel access for the same time-shares as achieved by single rate IEEE 802.11. This opportunistic mechanism is similar in principal to the design of proportional-fair scheduling algorithm [4] for 3G networks such as HDR (High Data Rate standard from Qualcomm).

8.4 Medium Access Layer

The original IEEE 802.11 [1] standard specifies the physical layer and the medium access layer mechanisms and provides a data rate up to 2 Mbps. The later standards IEEE 802.11b and IEEE 802.11a modifies the physical layer part of the standard and increases the maximum data rates to 11 Mbps and 54 Mbps respectively.

In this section, we first discuss the basic 802.11 MAC layer functionality called Distributed Coordination Function (DCF) for distributed access to the shared medium. We then discuss the Point Coordination Function (PCF) which provides a mechanism for centralized control of channel access. DCF is a natural choice for ad-hoc networks, as there is no centralized controller such as an access-point. However, PCF can support QoS metrics in single-hop wireless networks due to its centralized design. Both DCF and PCF are enhanced in the upcoming standard 802.11e [13] that is designed for supporting QoS in WLANs. We also present key features of the 802.11e protocol and discuss some service differentiation schemes that have been proposed for extending DCF.

8.4.1 802.11 Distributed Coordination Function (DCF)

The DCF protocol attempts to provide equal access (in terms of number of packets) to all backlogged nodes that share a channel. For example, in the Infrastructure mode if all nodes in a cell are in the transmission range of each other and there are no other sources of noise or interference, all users nodes and the AP get to send the same number of packets, assuming they all are backlogged.

In an ad-hoc network the throughput that a node obtains using DCF is a function of the number of neighbors that it has and the state of their queues (backlogged or not). Since the throughput of the neighbors depend on *their* neighbors, throughput determination becomes a global problem rather than a local problem. So, in general in an ad-hoc network using DCF the throughput received by a node depends on the whole topology. Note that the DCF mechanism attempts to provide access per-node and not per-link.

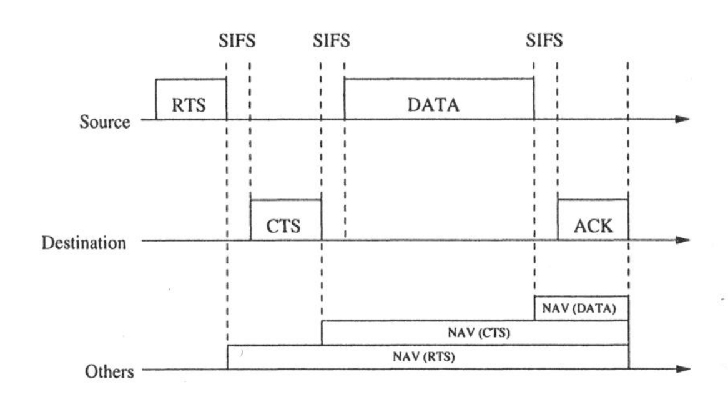

Figure 8.3. IEEE 802.11 DCF

We now describe the DCF mechanism is detail. Each node that has a packet to send picks a random slot for transmission in $[0, cw]$, where cw is the contention window used for backoffs. Initially cw is set to cw_{min}. In the chosen slot, the node sends a MAC layer control packet called RTS (request-to-send), to the receiver. If the receiver correctly receives the RTS and is not deferring transmission, it responds with a CTS (clear-to-send). This is followed by transmission of the data packet by the sender, and a subsequent acknowledgment from the receiver. The transmissions of these four packets are separated by short durations called SIFS (Short Inter-Frame Space). The SIFS allows time for switching the transceiver between sending and receiving modes. The sequence of transmission of these four packets is shown in Figure 8.3. The MAC header of all these packets (see the packet structures in Figure 8.4) contains a "duration" field indicating the remaining time till the end of the reception of the ACK packet. Based on this advertisement, the neighboring nodes update a data structure called NAV (Network Allocation Vector). This structure maintains the remaining time for which the node has to defer all transmissions.

If the packet transmission fails, the sender doubles its contention window ($cw \leftarrow [2 \times cw - 1]$) and backs off before attempting a retransmission. The number of retransmissions is limited to 4 for small packets (including RTS packets) and 7 for larger (typically DATA) packets. If these counts are exceeded, the data packet is dropped and cw is reset to cw_{min}. If the data packet is successfully delivered, both the sender and the receiver reset cw to cw_{min}.

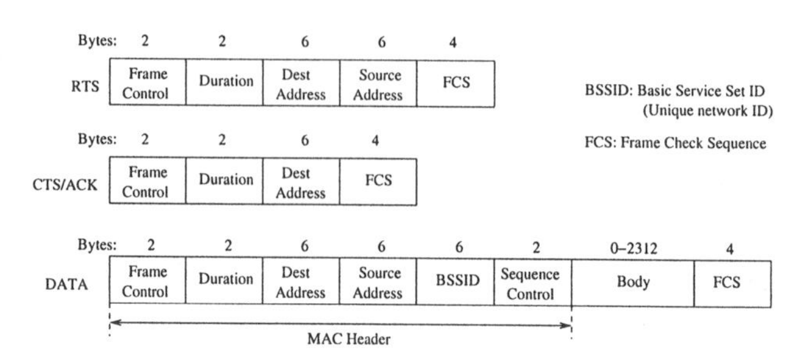

Figure 8.4. Packet formats for basic 802.11

8.4.2 802.11 Point Coordination Function (PCF)

PCF operates in the Infrastructure mode of 802.11. The standard requires that an AP implementing the PCF mode (contention-free period) must alternate it with the DCF mode (contention period). In the PCF mode, the point-coordinator (AP) sends packets to other nodes and polls a list of nodes giving them an opportunity to transmit. Unlike the DCF mode, in the PCF mode nodes can transmit only if they are polled by the AP. The beginning of the contention-free period (the period in which PCF operates), is marked by a beacon from the AP which also advertises the length of the contention-free period. During this period, the transmission schedule is completely determined by the AP. The contention-free period could be foreshortened by the AP by transmitting a special packet called the CF-End packet. The polls and the acknowledgments are piggybacked on the data packets as shown in Figure 8.5. Note that before sending the beacon, the AP waits for a period called the PIFS (PCF Inter Frame Space) which is larger than SIFS. This ensures that all communication related to the contention period has ceased. The PIFS interval is also used to wait for a response to a poll by the AP. After this interval elapses, the AP concludes that the node being polled either does not have packets to send or did not receive the poll. It then moves ahead by polling the next node after a PIFS period.

8.4.3 The QoS Extension: 802.11e

The IEEE 802.11e extension provides mechanisms for supporting different priorities in WLAN networks. Being a distributed protocol, it is hard to ensure strict priorities. Hence, the priorities are probabilistic in nature. Such priorities can be viewed as a form of QoS metric.

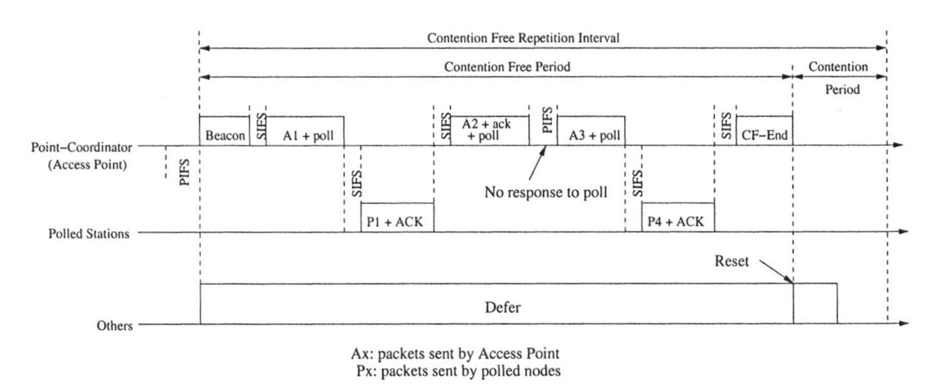

Figure 8.5. Point Coordination Function (PCF)

The DCF and PCF functionalities of 802.11 have been extended, and these extensions form the 802.11e standard[2]. The Enhanced DCF (EDCF) extends the functionality of DCF by providing the notion of priorities. The enhancement of PCF is called HCF (Hybrid Coordination Function) in 802.11e. Some of the mechanisms of 802.11e are similar to the service differentiation mechanisms to be discussed in Section 8.4.4.

Figure 8.6 shows the 802.11e functionality in detail.

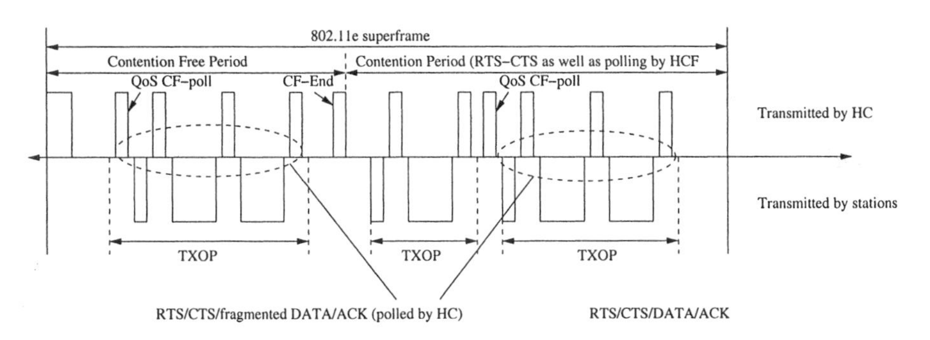

Figure 8.6. Example of a 802.11 super-frame. It relies on TXOPs (Transmission opportunities). Polled TXOP may be located in Contention Period or Contention-Free Period.

In EDCF, the frames entering the MAC layer can request 8 different service priorities. These priorities are mapped to different access categories (ACs). Each AC may have a distinct value for the DIFS period (now called AIFS), cw_{min} and cw_{max}. Figure 8.7 shows an example illustrating different class of

[2]The standardization is not yet complete

traffic with different AIFS values. These values can be dynamically determined by the access point. The nodes are informed of these values by using the beacon. The AIFS is at least as large as the DIFS in 802.11. Different priority levels will correspond to different values of AIFS.

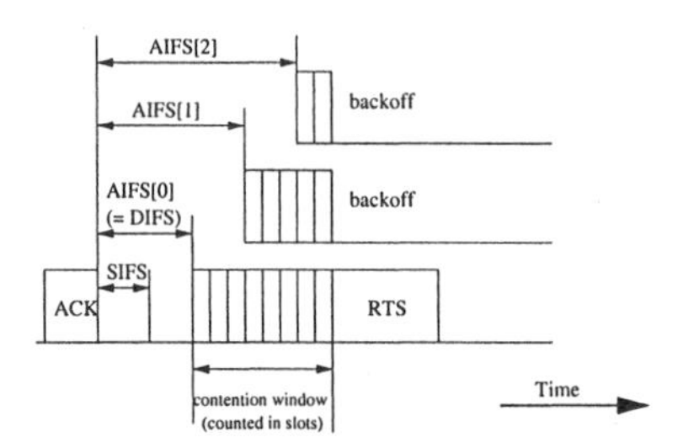

Figure 8.7. Multiple backoff of streams with different priorities

HCF allows the hybrid coordinator to maintain state for nodes and allocate contention free transmit opportunities (TXOP) in a smart way. The offered load per traffic class at each node is used by the hybrid coordinator for scheduling. Unlike in the case of the PCF mode of 802.11, the hybrid coordinator may poll user nodes in the contention-free period as well as in the contention-period.

Like the PCF in 802.11, this protocol requires centralized operation. To achieve the QoS requirements, the AP coordinates the transmissions in its cell. This protocol needs to be extended for ad-hoc networks where there is no centralized coordinator.

8.4.4 QoS Support using DCF based Service Differentiation

As it is difficult to provide absolute QoS guarantees, relative QoS assurance can be provided by service differentiation. This helps in designing systems which can support multiple classes of users.

As discussed in Section 8.4.1, in 802.11 all backlogged nodes contend for the channel using the same protocol with the same set of parameters. As a result, if all the contending nodes are in range of each other, 802.11 will provide long term fair share to each node. However, to provide differentiated services, the 802.11 protocol needs to be modified. [2] proposes three ways to modify the DCF functionality of 802.11 to support service differentiation. The parameters that need to be modified to achieve service differentiation are described below.

1 *Backoff increase function:* Upon an unsuccessful attempt to send an RTS or a data packet, the maximum backoff time is doubled. More specifically the backoff time is calculated as follows:

$$Backoff_{time} = \lfloor 2^{(2+i)} \times rand() \rfloor \times Slot_{time} \qquad (4.1)$$

where i is the number of consecutive backoffs experienced for the packet to be transmitted. To support different priorities, the backoff computation can be changed as follows

$$Backoff_{time} = \lfloor P_j^{(2+i)} \times rand() \rfloor \times Slot_{time} \qquad (4.2)$$

where P_j is the priority of node j.

2 *DIFS:* As shown in Figure 8.3, this is the minimum interval of time required before initiating a new packet transmission after the channel has been busy. To lower the priority of a flow we can increase the DIFS period for packets of that flow. However, it is difficult to find an exact relation between the DIFS period for a flow and its throughput. Figure 8.8 shows the different DIFS values and the corresponding relative priorities. This idea is similar to the concept of AIFS in 802.11e, as described in Section 8.4.3.

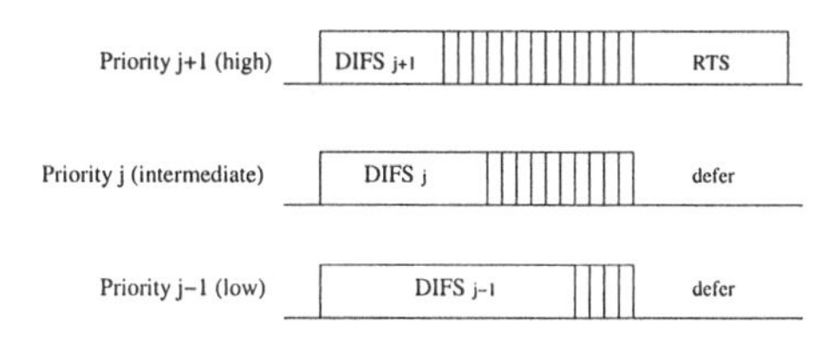

Figure 8.8. Service Differentiation using different DIFS values

3 *Maximum Frame Length:* Channel contention using the DCF functionality is typically used to send a single frame. By using longer frames, higher throughput can be provided to high-priority flows.

8.5 QoS Routing

The QoS metrics of an end-to-end route depends on the links of the computed route. There are three main challenges in computing a route satisfying QoS requirements. First, the QoS metric on each link must be either computed continuously or discovered on demand, when the route request packet is being forwarded. Second, broadcast based routing algorithms do not explore all possible routes. Third, mechanisms to compute the available bandwidth on a link are coarse and are based on observing other parameters such as queue length and channel access history.

Multi-hop networks are dynamic in nature, and transmissions are susceptible to fades, interference, and collisions from hidden/exposed stations. These characteristics make it a challenging task to design a QoS routing algorithm for multi-hop networks. Following are the main design goals for such an algorithm:

- The algorithm should be highly robust and should degrade gracefully with increasing mobility.
- Route computation should not require maintenance of global information.
- The computed route should be highly likely to sustain the requested bandwidth for the flow.
- The route computation should involve only a few hosts, as broadcast in the whole network is expensive.
- Hosts should have quick access to routes when connections need to be established.

AODV (Ad-hoc On-demand Distance Vector) and DSR (Dynamic Source Routing) are the first two routing protocols proposed for ad-hoc networks. Both the protocols are on-demand. AODV uses next-hop routing, whereas DSR uses source routing. More information on AODV and DSR can be found in [14]. A QoS routing protocol based on AODV for TDMA networks is proposed in [22]. An extension for DSR to support QoS is proposed in [11].

Rather than trying to fit QoS into the protocol, some routing protocols have been designed specifically for QoS routing. We describe two such protocols, namely CEDAR [16] and Ticket Based Routing [6, 20] in the remaining section.

8.5.1 Core Extraction based Distributed Ad-hoc Routing (CEDAR)

CEDAR achieves the above design goals for small to medium size ad-hoc networks consisting of tens to hundreds of nodes. The following is a brief description of the three key components of CEDAR.

- *Core Extraction:* A set of hosts is distributedly and dynamically elected to form the *core* of the network by approximating a minimum dominating set of the ad hoc network using only local computation and local state. Figure 8.9 shows an example network with four core nodes. Each core node maintains the local topology of the nodes in its domain, and also performs route computation on behalf of these nodes.

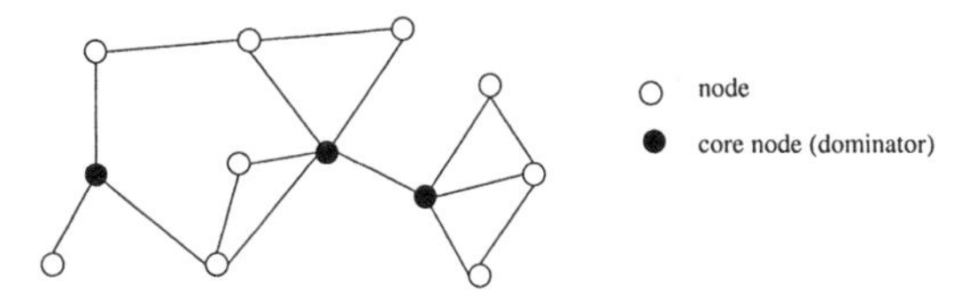

Figure 8.9. CEDAR: Core nodes in a network

- *Link state propagation:* QoS routing in CEDAR is achieved by propagating the bandwidth availability information of stable links in the core graph. The basic idea is that the information about stable high-bandwidth links can be made known to nodes far away in the network, while information about dynamic links or low bandwidth links should remain local.

- *Route Computation:* Route computation first establishes a core path from the dominator of the source to that of the destination. The core path provides the directionality of the route from the source to the destination. Using this directional information, CEDAR computes a route adjacent to the core path that satisfies the QoS requirements.

8.5.2 Ticket based routing

Ticket based routing [6] is based on the idea of limiting the broadcast messages and directing them toward the *right direction*. The goal of this approach is to select routes from the ones that are probed for route computation. The source has a certain number of tickets. Tickets are of two kinds: yellow and green. Each probe carries a certain number of tickets. The purpose of the yellow tickets is to maximize the probability of finding a feasible path. Hence probes carrying yellow tickets prefer paths with smaller delays. The purpose of the green tickets is to maximize the probability of finding a low-cost path, where each link is associated with a certain cost. Green tickets prefer paths with smaller costs, which may however have larger delay and hence have less chance to satisfy the delay requirement.

The source initiates the probing with a certain number of tickets of each color. At each intermediate node a decision is made as to how many tickets would be forwarded on each of the new probes. This decision is based on the observed QoS metrics of the link. For example, a link with lower delay gets higher number of yellow tickets compared to another link with higher delay.

The "Enhanced Ticket Based Routing Algorithm" approach [20] eliminates redundant probing and further optimizes ticket probing.

8.6 QoS at other Networking Layers

The need for QoS arises at the application layer. The application layer requests the transport layer to provide QoS services. The transport layer must request the routing layer to compute routes satisfying the QoS requirements. This request may need to travel down to the physical layer. Each layer receiving a QoS request from the above layer needs to take the following actions:

- *Check if it can be supported:* Each layer needs to see if the QoS requirements are within the limits of what it can support. It needs to notify the higher layer, if it can not support the QoS request.

- *Request the lower layer for supporting it:* The current layer processing the QoS request may be able to support it with the help of the lower layers. It needs to map the QoS requirement to the QoS services provided by the lower layer and then send the request to the lower layer. For example, for supporting a QoS route with a certain minimum bandwidth, the routing layer may inform the MAC layer to increase the priority of channel access.

- *Negotiate with the lower/upper layer:* When a QoS request is received from the upper layer, it should be checked if the network can support that request. If the QoS demands can not be met, a different QoS requirement may be negotiated by suggesting alternate values of the relevant QoS metrics.

- *Report the application layer on failure to support QoS:* After establishing a QoS connection, in case the network fails to support the QoS metrics, the application layer needs to be notified so that it can take appropriate actions. For example, if the network can not find routes requiring a certain minimum bandwidth for supporting real time communication, the application layer can change the encoding or resolution of the multimedia data. The networking layer noticing a change in observed QoS must report it up the layers to the application layer.

8.7 Inter-Layer Design Approaches

The previous sections discussed mechanisms at individual networking layers for providing QoS support in ad-hoc networks. The QoS support provided by a layer is dependent on the support from the lower layers as well. INSIGNIA [10] and Cross-Layer Design [5] are two efforts directed toward design and implementation of inter-layer QoS solutions. The rest of the section describes these two frameworks in detail.

8.7.1 INSIGNIA

In this framework the applications specify their minimum and maximum bandwidth needs. INSIGNIA is responsible for resource allocation, restoration control, and session adaptation between communicating mobile hosts. The design of the QoS routing protocol is independent of this framework.

This framework uses in-band signaling. There are two mechanisms that may be used for QoS related signaling: out-of-band and in-band. Out-of-band signaling refers to sending explicit control messages. In-band signaling refers to carrying control information as part of packet headers. Using in-band signaling flows/sessions can be rapidly established, restored, adapted, and released in response to wireless impairments and topology changes.

Various components of the architecture are shown in Figure 8.10. *Admission control* is responsible for allocating bandwidth to flows based on the maximum/minimum bandwidth requested. *Packet forwarding* classifies incoming packets and forwards them to the appropriate module (viz. routing, signaling, local applications, packet scheduling modules). *Routing* dynamically tracks changes in ad-hoc network topology, making the routing table visible to the node's packet forwarding engine. *Packet Scheduling* responds to location-dependent channel conditions when scheduling packets in wireless networks. *Medium Access Control (MAC)* provides quality of service driven access to the shared wireless media for adaptive and best effort services.

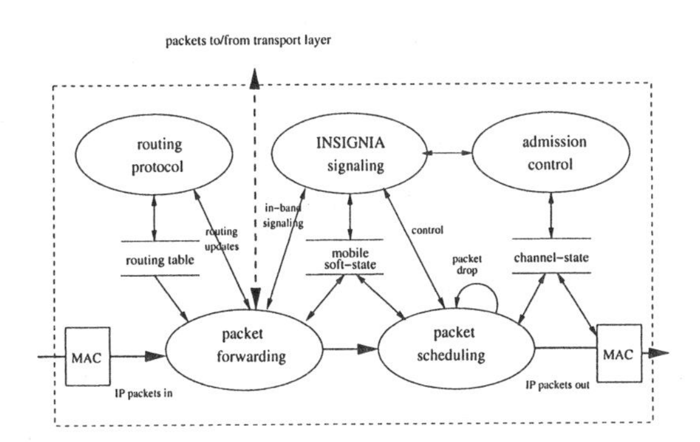

Figure 8.10. INSIGNIA QoS Framework

8.7.2 Cross-Layer Design for Data Accessibility

The architecture of the Cross-Layer Design [5] is shown in Figure 8.11. The application, middleware and the routing layers share information to achieve a higher quality in accessing data. The system relies on data replication to avoid the problem of missing data when network partitioning occurs. Map viewing and messaging are two examples shown in the figure.

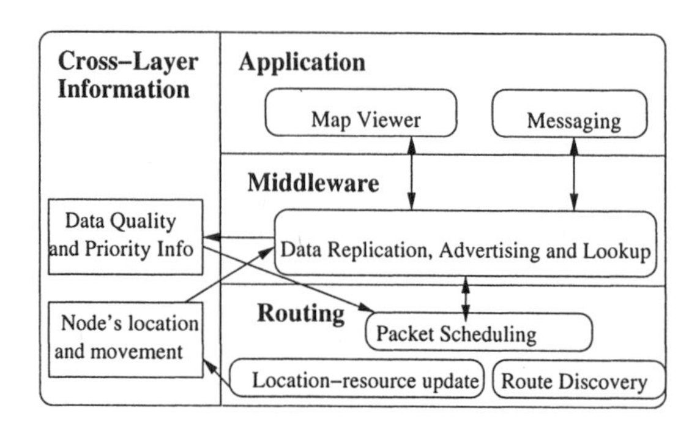

Figure 8.11. Cross-Layer Design for Data Accessibility

The routing layer uses a *predictive location-based routing protocol.* It uses each node's geometric coordinates and movement pattern information for the purpose of route discovery and maintenance. The *location-resource update* module periodically broadcasts messages containing the node's location and resource information to other nodes in the network. The routing layer reacts to route performance deterioration by route re-computation.

The middleware layer implements a *data accessibility service* that assists applications to advertise and share data with other users in the network. Data is accessed in two steps. In the first step, data availability information is obtained and presented to the application level. The QoS parameter of interest is the *success rate* in accessing data. In the second step the middleware layer retrieves the data from a remote host with certain application level requirements, such as data access deadline and data quality. The middleware layer translates the application level requirements into network level QoS parameters such as bandwidth and delay. It then sets up a route with these parameters. For sustaining QoS violations, the middleware layer is notified as the routing protocol will not be able to handle it. The middleware layer may adapt to the available QoS values.

8.8 Conclusion

In this chapter we studied QoS issues at various networking layers for ad-hoc networks. The physical layer and the MAC layers are primarily responsible for QoS on a single link. The DCF and PCF functionality of 802.11 is being extended into the QoS extension called 802.11e. The PCF and 802.11e protocols are specifically designed for QoS support in single-hop networks. These algorithms need to be adapted for use in multi-hop ad-hoc networks. The routing layer is responsible for computing and maintaining end-to-end multi-hop QoS routes. CEDAR [16] and Ticket Based Routing [20] protocols are two QoS routing protocols proposed for ad-hoc networks. Since the QoS needs arise at the application layer, the QoS requirements in the form of acceptable values for QoS metrics are specified by the application. The QoS request may have to travel down the network layers up to the physical layer. Applications would typically like to be notified in case the QoS requirements can not be met due to changes in the network conditions. The application may be able to (re)negotiate a different QoS requirement and adapt to it.

QoS is currently an active research area in ad-hoc networks. This chapter has covered some of the main research topics related to QoS in ad-hoc networks. However, there are several avenues that require further exploration for designing a QoS enabled ad-hoc network. We briefly outline some of these issues:

- *Energy efficient QoS architecture:* Ad-hoc networks are energy constrained as they are composed of hand-held devices with limited battery. Supporting QoS may require addition of extra in-band or out-of-band signaling messages, or other changes to protocols that increase the total energy needs. Hence, the QoS components of ad-hoc networks must be designed keeping energy efficiency as one of the key goals.

- *QoS metrics with level of tolerance:* The routing approaches such as CEDAR and the ticket based routing protocols attempt to compute QoS routes. These approaches do not provide hard guarantees on any QoS metric. The source can specify the amount of tolerance for each QoS metric and the network would then support the request based on the tolerance levels.

- *Multi-hop synchronized MAC Layer:* For packets that traverse multiple hops, the end-to-end QoS is a function of the QoS metrics at each intermediate link. End-to-end QoS properties can be improved by designing a MAC layer that coordinates with other intermediate nodes on a multi-hop path.

- *Extending PCF and 802.11e for Ad-hoc Networks:* Both the PCF and 802.11e solutions require the point coordinator (or the access point) to decide the transmission schedule. As there is no centralized control in an ad-hoc network, either this functionality needs to be performed distributedly or other changes need to be made to these protocols to use them in ad-hoc networks.

We find that QoS is an inherent component of ad-hoc networking and that there are several unsolved challenges that need to be addressed to design QoS enabled ad-hoc networks.

References

[1] IEEE Std. 802.11. Wireless LAN Medium Access Control (MAC) and Physical Layer (PHY) Specifications , 1999.

[2] I. Aad and C. Casellucccia. Differentiation mechanisms for ieee 802.11. In *Proc. IEEE Infocom*, volume 2, pages 594–602, 1996.

[3] G. Armitage. *Quality of Service in IP Networks.* Que; 1st edition, 2000.

[4] Paul Bender, Peter Black, Matthew Grob, Roberto Pad ovani, Nagabhushana Sindhushayana, and Andrew Viterbi. CDMA/HDR: A Bandwidth-Efficient High-Speed Wireless Data Se rvice for Nomadic Users. *IEEE Communications Magazine*, 38:70–77, Jul. 2000.

[5] K. Chen, S. H. Shah, and K. Nahrstedt. Cross Layer Design for Data Accessibility in Mobile Ad-hoc Networks. *Journal of Wireless Communications*, 21:49–75, 2002.

[6] S. Chen and K. Nahrstedt. Distributed QoS Routing with Imprecise State Information. In *Proc. of 7th IEEE International Conference on Computer, Communications and Networks (ICCCN'98)*, pages 614–621, October 1998.

[7] G. Holland, Nitin Vaidya, and Paramvir Bahl. A Rate-Adaptive MAC Protocol for Multi-Hop Wireless Networks. In *Proc. ACM MOBICOM*, June 2001.

[8] S. Jiang, D. He, and J. Rao. A prediction-based link availability estimation for mobile ad-hoc networks. In *Proc. IEEE Infocom*, pages 1745–1752, 2001.

[9] A. Kamerman and L. Monteban. WaveLAN II: A High-performance Wireless LAN for the Unlicensed Band. *Bell Labs Technical Journal*, pages 118–133, Summer 1997.

[10] S.-B. Lee, G.-S. Ahn, X. Zhang, and A. T. Campbell. INSIGNIA: An IP-Based Quality of Service Framework for Mobile Ad-hoc Networks. *Journal of Parallel and Distributed Computing*, 60:374–406, 2000.

[11] R. Leung, J. Lio, E. Poon, A.-L. C. Chan, and B. Li. Mp-dsr: A qos-aware multi-path dynamic source routing protocol for wireless ad-hoc networks. In *Proc. IEEE Conference on Local Computer Networks (LCN'01)*, 2001.

[12] C. R. Lin and J. S. Liu. QoS Routing in Ad-hoc Wireless Networks. *IEEE Journal on Selected Areas in Communications*, 17(8), 1999.

[13] S. Mangold, S. Choi, G. R. Hiertz, O. Klein, and B. Walke. Analysis of IEEE 802.11e for QoS Support in Wireless LANs . *IEEE Wireless Communications Magazine, Special Issue on Evolution of Wireless LANs and PANs*, Jul. 2003.

[14] C. E. Perkins. *Ad Hoc Networking*. Addison-Wesley, Reading, MA, 2001.

[15] B. Sadeghi, V. Kanodia, A. Sabharwal, and E. Knightly. Opportunistic Media Access for Multirate Ad-hoc Networks. In *Proc. ACM MOBICOM*, 2002.

[16] P. Sinha, R. Sivakumar, and V. Bharghavan. Cedar: a core-extraction distributed ad hoc routing algorithm. In *Proc. IEEE Infocom*, 1999.

[17] Jungmin So, , and N. H. Vaidya. A Multi-channel MAC Protocol for Ad-hoc Wireless Networks. In *Technical Report*, January 2003.

[18] A. Veres, A. T. Campbell, M. Barry, and L. H. Sun. Supporting service differentiation in wireless packet networks using distributed control. *IEEE Journal on Selected Areas in Communications*, October 2001.

[19] Xi'an. Dynamic tdma slot assignment in ad-hoc networks. In *Proc. 17th International Conference on Advanced Information Networking and Applications (AINA'03)*, 2003.

[20] L. Xiao, J. Wang, and K. Nahrstedt. The Enhanced Ticket Based Routing Algorithm. In *Proc. IEEE ICC*, 2002.

[21] G. G. Xie and J. Gibson. A Networking Protocol for Underwater Acoustic Networks. Technical Report, TR-CS-00-2, Dept. of Computer Science, Naval Postgraduate School, December 2000.

[22] C. Zhu and M. S. Corson. Qos routing for mobile ad-hoc networks. In *Proc. IEEE Infocom*, 2002.

Chapter 9

SECURITY IN MOBILE AD-HOC NETWORKS

Yongguang Zhang
HRL Laboratories, LLC

Wenke Lee
College of Computing, Georgia Institute of Technology

Abstract Security is a paramount concern in mobile ad hoc network (MANET) because of its intrinsic vulnerabilities. These vulnerabilities are nature of MANET structure that cannot be removed. As a result, attacks with malicious intent have been and will be devised to exploit these vulnerabilities and to cripple MANET operations. In this chapter, we analyze the security problems in MANET and present a few promising research directions. On the prevention side, various key and trust management schemes have been developed to prevent external attacks from outsiders, and various secure MANET routing protocols have been proposed to prevent internal attacks originated from within the MANET system. On the intrusion detection side, a new intrusion detection framework has been studied especially for MANET. Both prevention and detection methods will work together to address the security concerns in MANET.

Keywords: Vulnerabilities, Attack prevention, Key management, Secure routing, Intrution dection

9.1 Vulnerabilities of Mobile Ad Hoc Networks

Security is a paramount concern to mobile ad hoc networking (MANET) because a MANET system is much more vulnerable to malicious exploits than a wired (traditional) network. First of all, the use of wireless links renders the network susceptible to attacks ranging from passive eavesdropping to active interfering. Unlike wired networks where an adversary must gain physical access to the network wires or pass through several lines of defense at firewalls and gateways, attacks on a wireless network can come from all directions and target at any node. Damages can include leaking secret information, message contamination, and node impersonation. All these mean that a wireless ad-hoc network will not have a clear line of defense, and every node must be prepared for encounters with an adversary directly or indirectly.

Second, mobile nodes are autonomous units that are capable of roaming independently. This means that nodes with inadequate physical protection are receptive to being captured, compromised, and hijacked. Since tracking down a particular mobile node in a large scale ad hoc networks may not be easily done, attacks by a compromised node from within the network are far more damaging and much harder to detect. Therefore, mobile nodes and the infrastructure must be prepared to operate in a mode that trusts no peer.

Third, decision-making in mobile computing environment is sometimes decentralized and some wireless network algorithms rely on the cooperative participation of all nodes and the infrastructure. The lack of centralized authority means that the adversaries can exploit this vulnerability for new types of attacks designed to break the cooperative algorithms.

For example, many of the current MAC protocols for wireless channel access are vulnerable. Although there are many types of MAC protocols, the basic working principles are similar. In a contention-based method, each node must compete for control of the transmission channel each time it sends a message. Nodes must strictly follow the pre-defined procedure to avoid collisions and to recover from them. In a contention-free method, each node must seek from all other nodes a unanimous promise of an exclusive use of the channel resource, on a one-time or recurring basis. Regardless of the type of MAC protocol, if a node behaves maliciously, the MAC protocol can break down in a scenario resembling a denial-of-service attack. Although such attacks are rare in wired networks because the physical networks and the MAC layer are isolated from the outside world by layer-3 gateways/firewalls, every mobile node is completely vulnerable in the wireless open medium.

Furthermore, mobile computing has introduced new type of computational and communication activities that seldom appear in fixed or wired environment. For example, mobile users tend to be stingy about communication due to slower links, limited bandwidth, higher cost, and battery power constraints; mechanisms like disconnected operations [38] and location-dependent operations only appear to mobile wireless environment. Unsurprisingly, security measures developed for wired network are likely inept to attacks that exploit these new applications.

Applications and services in a mobile wireless network can be a weak link as well. In these networks, there are often proxies and software agents running in base-stations and intermediate nodes to achieve performance gains through caching, content transcoding, or traffic shaping, etc. Potential attacks may target these proxies or agents to gain sensitive information or to mount DoS attacks, such as flushing the cache with bogus references, or having the content transcoder do useless and expensive computation.

To summarize, a mobile ad-hoc network is very receptive to security attacks due to its open medium, dynamically changing network topology, cooperative algorithms, lack of centralized monitoring and management point, and lack of a clear line of defense. These vulnerabilities are nature of MANET structure that cannot be removed. As a result, attacks with malicious intent have been and will be devised to exploit these vulnerabilities and to cripple MANET operations.

9.2 Potential Attacks

We next look at a few attacks that are designed to exploit the vulnerabilities of MANET. We can often classify such attacks into external attacks, in which the attacker is in the proximity but not a trusted node in the ad-hoc network, and internal attacks, where the attackers are actually willing participants in the ad-hoc networks. External attacks are usually contained with conventional security mechanisms like membership authentication. This however cannot curb internal attacks because the issue here is not the identity but the behavior of the malicious participant. Since many ad-hoc network applications have the elements of spontaneous and open networking, internal attacks by participants are far more difficult to prevent. Further, external attacks are often mere stepping-stones leading to internal attacks, when an outside attacker gains total control of one (any) ad-hoc network node. Therefore, we here mainly focus on internal attacks and assume that attackers are actually network participants.

We also focus on the ad-hoc routing layer because that is a foundation for the ad-hoc network operations. Lessons learned in routing attacks can be applied to other layers and the whole systems as well.

Attacks on routing layer can be grossly classified into two categories, attacks on routing protocols and attacks on packet forwarding/delivery. Attacks on routing are designed to prevent a victim from knowing the path to a destination even if such a path exists in the network. Attacks on forwarding is to disrupt the packet delivery along a predetermined path.

Attacks on routing protocols can create various undesirable effects that defeat the objectives of ad-hoc routing, such as *network partition*, *routing loop*, *resource deprivation* (forcing all routes to pass through a victim), or *route hijack* (forcing all routes to pass through a malicious node). There can be many types of such attacks. They usually involve disseminations of false routing information (Route Request, Route Reply, Route Error, etc.), such as altering the path, falsifying the metric or sequence number, or fabricating untrue reports, etc. Here are several examples that have been studied in the literature [30] [18] [37] [17]:

- Impersonating another node to spoof route message.

- Advertising a false route metric to misrepresent the topology.
- Sending a route message with wrong sequence number to suppress other legitimate route messages.
- Flooding Route Discover excessively as a denial-of-service (for both routing and forwarding).
- Modifying a Route Reply message to inject a false route.
- Generating bogus Route Error to disrupt a working route.
- Suppressing Route Error to mislead others.

Preventing these attacks will be difficult. Authentication of message source is of limited use (except the first case above) because the source is often a legitimate (albeit malicious) participant node. Due to constant mobility and dynamic changing topology, it is very difficult to independently validate each route message. Some proposed solutions attempt to cryptographically tie the routing information or bound its ranges to the node' positions, neighborhood, and the exact path it has traversed, etc. These approaches however comes with added overhead and complexity and requires mechanisms specific to individual protocols.

There are also more sophisticated routing attacks. Compared to the simple attacks described above, these sophisticated attacks are much harder to detect and to prevent:

- *Wormhole attacks* [19]: two collaborating malicious nodes create a tunnel (virtual link) between them to falsify the widely-used hop-count metric.
- *Rushing attacks* [20]: to target certain routing protocols that choose routes on what message arrives first – a rushed malicious route message may block legitimate messages that arrive later.
- *Sybil attacks* [12]: one malicious node takes up multiple identities to project a false topology.

Even if we do have secured ad-hoc routing, attacks on forwarding can still disrupt the packet delivery. These attacks achieve two main goals: selfishness and denial-of-service. In a selfishness scenario, a malicious participant selectively drops data packets that it is supposed to forward in order to save its own resource. In a denial-of-service scenario, a malicious node can send excessive traffic through a victim node in order to deprive its battery power. There haven't been an easy way to detect and prevent these types of attacks.

9.3 Attack Prevention Techniques

Attack *prevention* measures, such as authentication and encryption, can be used as the first line of defense to reduce the possibilities of attacks. Most of the security research efforts in MANET to date, e.g., [40] [22] [45] [34] [4] [3] [42] [30] [18], are on attack prevention techniques. For example, (session) shared secret key schemes can be used to encrypt messages to ensure the confidentiality, and to some degree the authenticity (group membership), of routing information and data packets; more elaborate public key schemes can be employed to sign and encrypt messages to ensure the authenticity (of individual nodes), confidentiality, and non-repudiation of the communications between mobile nodes. The prevention schemes proposed so far differ in several ways, depending on their assumptions on the intended MANET applications.

9.3.1 Key and Trust Management: Preventing External Attacks

Encryption, authentication, and key management are widely used to prevent external (outsider) attacks. They however face many challenges in ad-hoc networks. First, we must deal with the dynamic topologies, both in communications and in trust relationship; the assessment of whether to trust a wireless node may change over time. Second, we must deal with the lack of infrastructure support in MANET; any centralized scheme may face difficulties in deployment.

Both symmetric and asymmetric key systems have been proposed for MANET. Some schemes, e.g., [3], use secret key encryption for efficiency and simply assume group membership is a sufficient authentication. Such symmetric key systems have the performance advantages but scalability disadvantages: it requires $O(n^2)$ keys (one between any pair of nodes). Blom's scheme [5] can reduce the space requirement down to a pre-selected threshold t where t can be much smaller than n, but it is only resilient against node capture up to $t-1$ nodes. Others use public keys, e.g., [45], to prevent routing messages being falsified by compromised nodes. Such asymmetric key signing are unfortunately very expensive.

Key generation, distribution and management in MANET is challenging because of the absence of central management. it is widely recognized that a MANET cannot rely upon centralized trust entities like key distribution center (KDC) and certificate authority (CA). Instead, distributed key management services such as those based on secret sharing scheme should be used [45] [25]. In a secret sharing scheme, the signing key for the otherwise centralized CA is distributed among n nodes using m-out-of-n threshold cryptography scheme: any m nodes can join together to provide the CA service, but not for any smaller set of nodes. This way, the distributed CA can tolerate up to $m-1$ compromises.

Another approach to the key distribution problem is to use a PGP-like "web of trust" certificates scheme to bootstrap trust relationships without using a trusted PKI [21]. Here, each node issues certificates for nodes that it trusts and builds subgraphs for in-bound and out-bound trusts (assuming trust is transitive as in PGP). For two nodes to establish secure communication, they merge their subgraphs to see if they intersect.

It is also a difficult problem establishing the initial trust base among nodes in a MANET. Without a pre-defined trust relationship, nodes of a spontaneous and open ad-hoc network will appear as "strangers" to each other [2]. Stajano and Anderson [41] proposed a scheme that establishes secure transient association between mobile devices by "imprinting" according to the analogy to duckling acknowledging the first moving subject they see as their mother. Eschenauer and Gligor [14] developed a random key pre-distribution scheme, where each node chooses a set of k key randomly from a large common key pool P. The size of P and k are properly chosen so that the probability of any two nodes sharing a least one key is very high. Chan el at. [9] further improved the security of this scheme by requiring two nodes to share q keys instead of one. Further, key agreement protocols (such as [13]) can be constructed based on Blom's scheme [5] to pre-distribute keys to MANET nodes.

9.3.2 Secure Routing Protocols: Preventing Internal Attacks

The above mechanisms are useful to authenticate MANET nodes and prevent outsiders from masquerading as internal nodes. They however cannot prevent internal attacks such as misbehaving nodes attacking on ad-hoc routing. This will require secure routing with hardened protocols that force every nodes to abide the rules. Indeed, several such secure MANET routing protocols have been proposed to enhance or replace existing ones. For example, SEAD [17] (Secure Efficient Ad Hoc Distance Vector) has been proposed to replace DSDV [32] as a secure distance-vector-based MANET routing protocol. Ariadne [18], a new secure on-demand ad-hoc routing protocol, can secure DSR [24] and prevent its most severe attacks such as modifying the discovered routes. Two new protocols, ARAN [37] and SAODV [42], have been proposed to secure AODV [33] with public key cryptography. Finally, Papadimitratos et al. have proposed a completely different routing protocol called SRP (Secure Routing Protocol) [30].

In these secure routing protocols, cryptographic mechanisms are widely used to protect the routing messages, or to provide a mean to prove their bounds. For example, metric authentication can be built with one-way hash chain, as in SEAD, to authenticate metric and sequence number. The metric increase can be tied to the forwarding node so that an attacker cannot replay a heard metric or fabricate one. Route authentication can be built with pairwise shared key as described in Ariadne [18]: the i-th node in the path forwarding a route request packet p computes a MAC using the key K_{iD} it shares with target D and includes a hash $h_i = H(N_i||\text{MAC}(K_{iD}, p)||h_{i-1})$ in the message – the target can then verify all the nodes in the path. The cost of using pairwise shared keys can be further reduced by adopting TELSA [35], an efficient broadcast authentication protocol that requires loose time synchronization, to secure route discovery and maintenance [18]. To use TELSA for authentication, a sender generates a hash chain and determines a schedule to publish the keys of the hash chain. The key required to authenticate a packet will not be published before the packet has arrived at the receiver so that an adversary cannot have captured the key and forged the packet.

Some secure routing protocols such as SRP [30] use the multipath feature of an ad-hoc network to achieve better security properties. SRP uses a Secure Message Transmission (SMT) protocol to disperse a message into n pieces and transmits them through different paths, given a topology map of the network. A successful reception of any m-out-of-n pieces allows the reconstruction of the original message.

9.3.3 Limitations of Prevention Techniques

While the techniques discussed in this section can prevent and deter certain attacks in MANET, there is a limitation to the effects of prevention techniques in general. First, these techniques are designed for a set of *known* attacks. They are unlikely to prevent newer attacks that are designed to circumvent the existing security measures. We must have a second mechanism to detect these newer attacks. Second, each of the prevention techniques comes with added overhead and complexity. Given the resource constraints in MANET, it is not realistic to have all known prevention techniques activated at all time. Unfortunately, we do not yet have a good understanding of the resource consumption characteristics of the prevention techniques, and nor have we developed good strategies for activating the appropriate mechanisms according to run-time conditions.

Experience in security research in the wired environments has taught us that we need to deploy defense-in-depth or layered security mechanisms because security is a process (or a chain) that is as secure as its weakest link [39]. In addition to prevention, we also need *detection* and *response*, as well as security policies and vulnerability analysis. Clearly, the same principle applies to MANET because not a single approach can solve all MANET security problems.

9.4 Intrusion Detection Techniques

Although much attention in building a secure mobile ad-hoc network is still focused on prevention techniques as shown in the previous section, researchers have begun to investigate detection and response schemes as well. Partridge et al. [31] report that basic signal processing techniques can be used to perform traffic analysis on packet streams, even if the data is encrypted. Marti et al. [28] propose to use "watchdog" to identify nodes with routing misbehavior and to avoid such nodes in the route used. It also uses "pathrater" to choose better path based on the reputation of intermediate nodes if multiple paths are available. CONFIDANT [8] further extends these approaches to evaluate the level of trust of alert reports and to include a reputation system to rate each node. Hsin et al. [16] study a static sensor network and propose a power-efficient distributed neighbor monitoring mechanism where alarms are transmitted back to a control center. Bucegger et al. [7] propose a routing protocol extension that detects and isolates nodes that do not cooperate in routing and forwarding due to selfishness. Finally, Zhang et al. [43] were the first to discuss the need for a general intrusion detection framework in MANET. A follow-up work [44] focuses on a preliminary investigation of anomaly detection approaches for MANET.

The primary assumptions of intrusion detection are: user and program activities are observable, for example via system auditing mechanisms; and more importantly, normal and intrusion activities have distinct behavior. Intrusion detection therefore involves capturing audit data and reasoning about the evidence in the data to determine whether the system is under attack. Based on the type of audit data used, intrusion detection systems (IDSs) can be categorized as network-based or host-based. A network-based IDS normally runs at the gateway of a network and "captures" and examines network packets that go through the network hardware interface. A host-based IDS relies on operating system audit data to monitor and analyze the events generated by programs or users on the host. The same methodology can be applied to intrusion detection in MANET, but it must be adapted to the new environment and new requirements.

9.4.1 Architecture Overview

Due to the dynamic nature of MANET, intrusion detection and response in MANET must be distributed and cooperative [43]. In this architecture, as shown in Figure 9.1, "monitoring nodes" throughout the network each runs an IDS agent. In the "every node" scheme, every node can be the monitoring node for itself. Alternatively a "clustering-based" scheme can be derived for better efficiency, where a cluster of neighboring nodes can elect a node to be the monitoring node for the neighborhood.

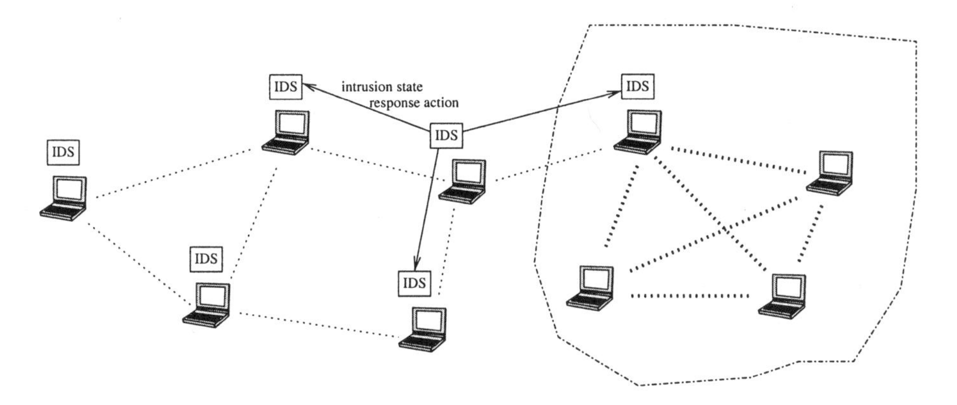

Figure 9.1. An IDS architecture for mobile ad-hoc network: IDS agents run on monitoring nodes throughout the network. Each MANET node can be the monitoring node for itself. Alternatively, a cluster of neighboring nodes can share one monitoring node.

Each IDS agent runs independently and is responsible for detecting intrusions to the local node or its cluster. IDS agents on neighboring monitoring nodes can collaboratively investigate to not only reduce the chances of producing false alarms, but also detect intrusions that affect the whole or a part of the network. These individual IDS agents collectively form the IDS to defend the MANET. The internal of an IDS agent, as shown in Figure 9.2, can be conceptually structured into six pieces: the data collection module, the local detection engine, the cooperative detection engine, the local response and global response modules, and a secure communication module that provides a high-confidence communication channel among IDS agents.

Data Collection. The local data collection gathers streams of real-time audit data from various sources. Useful data streams can include system and user application data, network routing and data traffic measurements, as well as activities observable within the radio range of the monitoring node. Multiple data collection modules can coexist in one IDS agents to provide multiple audit streams for a multi-layer integrated intrusion detection method.

Local Detection. The local detection engine analyzes the local data traces gathered by the local data collection module. It can use both misuse and anomaly detection algorithms. It is likely that the number of newly created attack types mounted on mobile computing environments will increase quickly as more and more network appliances become mobile and wireless. It is therefore very important that we focus more on anomaly detection techniques. We will present a preliminary case study of anomaly detection for ad-hoc routing protocols later in Section 9.4.3.

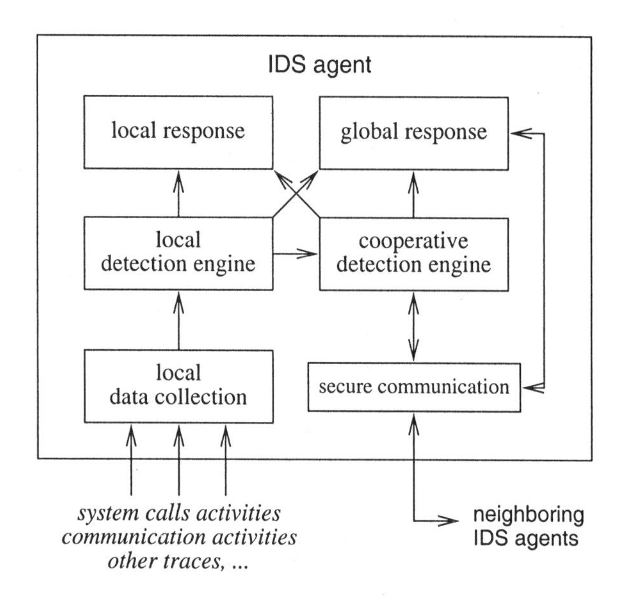

Figure 9.2. A conceptual model of an IDS agent

Regardless of the detection methods (i.e., misuse or anomaly detection) used in a MANET, we need to address the issue of how to systematically build ID models that are both effective and efficient. This will be discussed later in Section 9.4.2.

Cooperative Detection. An IDS agent that detects locally a known intrusion or anomaly with *strong* evidence (i.e., the detection rule triggered has a very high accuracy rate) can determine independently that the network is under attack and can initiate a response. However, if a node detects an anomaly or intrusion with weak evidence, or the evidence is inconclusive but warrants broader investigation, it can initiate a cooperative global intrusion detection procedure. This procedure works by propagating the intrusion detection state information among neighboring agents. If an agent(s), using alert information from other agents, now finds the intrusion evidence to be sufficiently strong, it initiates a response.

We consider cooperative detection a problem similar to local detection because both look for evidence of intrusion or anomaly using the information gathered. The difference is that in local detection an IDS agent collects and analyzes information about the local node (or the cluster), whereas in cooperative detection the IDS agent relies on alert data from other IDS agents. Therefore, the learning-based approach for building ID models (Section 9.4.2 below) can be applied to both local and cooperative detection.

Cooperative detection can result in lower false alarm rate because local intrusion report can be confirmed by others. It can also helps the investigation and identification the compromised node(s) behind the intrusion. For example, routing "blackhole" and network "partitioning" attacks usually result in anomalies observable by multiple IDS agents, which can then share the information to pinpoint the likely compromised node(s).

Local and Global Response. Intrusion response in MANET depends on the type of intrusion, the help (if any) from other security mechanisms, and the application-specific policy. An example response is to re-authenticate the nodes and re-organize the network, e.g., by re-initializing communication channels between the re-authenticated legitimate nodes, to exclude the compromised node(s).

9.4.2 A Learning-Based Approach

Intrusion detection in MANET is a very challenging task because there are many potential (and new) attacks, and because the distinction between intrusions and legitimate conditions is not always obvious due to the dynamically changing topology and volatile physical environment. In order to be effective (i.e., highly accurate), an ID model must perform comprehensive analysis on an extensive set of features.

One way to build such ID models is to use a learning-based approach for automatically selecting and constructing appropriate features from audit data and computing ID models. The main idea is to first start with a (broad) set of features, perhaps enumerated using domain knowledge, then apply data mining algorithms (e.g., [1] [27]) to compute temporal and statistical patterns describing the correlations among the features and the co-occurring events. The consistent patterns of normal activities and the unique patterns associated with intrusions are then identified and analyzed to select the appropriate features or construct additional features. Machine learning algorithms [29] (e.g., the RIPPER [11] classification rule learner) are then used to compute the detection models.

In this approach, the selected and constructed features are seeded from domain knowledge but are more empirical and objective because they are based on patterns computed from audit data. The inductively learned ID rules are usually more generalizable than hand-coded rules. That is, they tend to have better performance against new variants of known normal behavior or intrusions. This is because when there is more than one candidate model, classification algorithms always produce the model with better performance on a hold-out dataset, which is not used to produce the models and is intended to simulate the situation of encountering unseen or future cases.

The learning-based approach toward ID models has been proved successful in wired network environment [26]. It is therefore rational to believe that this approach, complemented with the use of expert knowledge, can achieve the objective of providing systematic tools for IDS developers to construct ID models quickly and easily for MANET. However, it is also conceivable to expect that this approach will face the following new challenges as MANET has introduced new constraints and new requirements to ID models.

Multitude of ID models. It is impractical to compute (or train) ID models on-line with MANET nodes because of the resource constraints. Instead, the models need to be computed off-line using simulation or historical data. Therefore, we may have to train a wide range of ID models each suitable for a class of similar application scenarios. At run-time, we will attempt to identify the run-time scenario and activate the appropriate model (if one is indeed available).

Focus on anomaly detection. Unlike the case of misuse detection, where a machine learning algorithm is given a set of data labeled with normal or intrusions to compute a classifier as an ID model, often there is only normal data available for training an anomaly detection model. Anomaly detection assumes that strong feature correlation exists in normal behavior, and such correlation can be used to detect deviations caused by abnormal (or intrusive) activities. Therefore, we should use a cross-feature analysis approach that explores correlations between each feature and all other features. This approach computes a classifier for each f_i using $\{f_1, f_2, \ldots, f_{i-1}, f_{i+1}, \ldots, f_L\}$, where $\{f_1, f_2, \ldots f_L\}$ is the feature set. The original anomaly detection problem, i.e., whether a record described by the feature set is normal or not, is then transformed into a set of sub-problems each examining whether the actual value of the feature matches with what the corresponding classifier has predicted. A mismatch is assigned a score according to the confidence (i.e., its accuracy in training) of the classifier. The scores are then combined to generate a final anomaly score. If it is above a threshold, then the original record is deemed anomalous.

Learning-based approach for cooperative detection. The learning-based approach needs to be applied to build a cooperative detection model. Our idea is to compute a classifier for A_0 using $\{A_1, A_2, \ldots, A_{n-1}\}$ as features, where A_i is the alert from node i. The computed detection model is to be used on node 0. That is, the classifier models the correlation between the alert by the local ID agent and alerts from neighboring ID agents. n, the number of ID agents that participate in cooperative detection, is obviously not fixed in run-time. Our approach is to (in the training phase) determine the minimum n *required* to produce a sufficiently accurate model. In run-time, if there are more than n agents participating, alerts from the *top* n agents (e.g., that are the closest in distance) are used. Otherwise, the model cannot be used, and no cooperative detection and response will take place (at this node).

9.4.3 Case Study: Anomaly Detection for Ad-Hoc Routing Protocols

In this section, we present a preliminary study to illustrate the research issues and our proposed approaches outlined earlier. Although we currently focus on the ad-hoc routing protocols, and intrusion detection at different network layers may use different audit data and have different performance and efficiency requirements, we believe that the same principles apply to the problems of building ID models for other layers.

Our objective in this study is to lead to a better understanding of the important and challenging issues in intrusion detection for ad-hoc routing protocols. First, we want to identify which routing protocol, with potentially all its routing table information used, can result in better performing detection models. This will help answer the question "what information should be included in the routing table to make intrusion detection effective." This finding can be used to design more robust routing protocols. Next, using a given routing protocol, we can explore the feature space and algorithm space to find the best performing model. This will give insight to the general practices of building intrusion detection for mobile networks.

MANET Environments. We choose two specific wireless ad-hoc protocols as the subject of our study. They are Dynamic Source Routing (DSR) Protocol [24] and Ad-hoc On-demand Distance Vector (AODV) Protocol [33]. There are other MANET routing protocols such as ZRP [15], OLSR [10], etc. We consider the above two protocols because they have been intensively studied in recent research. They have competitive performance under high load and mobility. We used the wireless network simulation software from network simulator ns2 [6] (release 2.1b9a, July 2002) in our study. It includes simulation for wireless ad-hoc network infrastructure, popular wireless ad-hoc routing protocols (e.g., DSR, AODV, and others), and mobility scenario and traffic pattern generation.

Attack Models. In our study, we implemented the following attacks in simulation: (1) blackhole attack where a malicious node advertise itself as having the shortest path to all nodes in the environment; and (2) selective packet dropping where a malicious node drops packets based on packet destinations or some other characteristics. The first is representative routing attack and the second is an attack on packet forwarding.

Audit Data. We suggest these three *local* data sources be used for anomaly detection: (1) topology information, such as node moving speed, (2) local routing information, such as route cache entries and route updates; and (3) traffic information, all incoming and outgoing traffic statistics, including inter-arrival periods and frequencies. We use only local information because remote nodes can be compromised and their data cannot be trusted. The intuition here is that there should be correlation between node movements and routing table changes, and between routing changes and traffic changes under normal conditions, and that such information can be used to detect anomalies caused by attacks.

Features. In our study, we define a total of 141 features according to domain knowledge and intuition. These features belong to two categories, non-traffic related and traffic related. The non-traffic related features are listed in Table 9.1. They capture the basic view of network topology and routing operations. In addition, "absolute velocity" characterizes the physical movement of a node.

The traffic related features are collected based on the following considerations. Packets come from different layers and different sources. For example, it can be a TCP data packet or a route control message packet (for instance, a `ROUTE REQUEST` message used in AODV and DSR) that is being forwarded at the observed node. We can define the first two aspects or dimensions of a traffic feature as, *packet type*, which can be data specific and route specific (including different route messages used in AODV and DSR), and *flow direction*, which can take one of the following values, `received` (observed at destinations), `sent` (observed at sources), `forwarded` (observed at intermediate routers) or `dropped` (observed at routers where no route is available for the packet). We need to evaluate both short-term and long-term traffic patterns. In our experiments, we sample data in three predetermined *sampling periods*, 5 seconds, 1 minute and 15 minutes. Finally, for each traffic pattern, we choose two typical *statistics measures* widely used in literature, namely, the packet count and the standard deviation of inter-packet intervals. Overall, a traffic feature has the following dimensions: packet type, flow direction, sampling periods, and statistics measures. An example is the feature that computes the standard deviation of inter-packet intervals of received `ROUTE REQUEST` packets every 5 seconds. All dimensions and allowed values for each dimension are defined in Table 9.2.

Classifiers. We use the cross-feature analysis approach discussed in Section 9.4.2 for anomaly detection, where a classifier is built for each feature using the rest of the features. We use several classification algorithms for evaluation purposes. These classifiers are C4.5 [36], a decision tree classifier, RIPPER [11], a rule based classifier, SVM Light [23], a support vector machine

Table 9.1. Feature Set I: Topology and route related features

Features	Notes
Time	ignored in classification, only used for reference
Absolute velocity	
Route add count	Routes newly added by route discovery
Route removal count	Stale routes being removed
Route find count	Routes found in cache and do not need to re-discovery
Route notice count	
Route repair count	Broken routes currently under repair
Total route change	
Average route length	

Table 9.2. Feature Set II: Traffic related features: dimensions and allowable values

Dimension	Values
Packet type	data, route, `ROUTE REQUEST`, `ROUTE REPLY`, `ROUTE ERROR`, and `HELLO` messages
Flow direction	received, sent, forwarded, and dropped
Sampling periods	5, 60 and 900 seconds
Statistics measures	count and standard deviation of inter-packet intervals

classifier. and NBC, a naive Bayes classifier. These inductive classifiers are all very efficient, which is important for MANET.

Effectiveness. We use trace data of normal runs for training the anomaly detection models. We then run the attacks and collect the trace data for evaluating the models. For example, if in a simulation the MANET total running time is 10,000 seconds, and the sample rate, by which the feature values are computed, is 5 seconds, then the trace data has 2,000 data points or *events*. Each event is labeled as normal or abnormal according to when and for how long an attack is run (and how long the effect lasts). When evaluating an anomaly detection model, we compute how many abnormal events are correctly identified (i.e., the detection rate) and how many normal events are incorrectly identified as anomalies (i.e., the false alarm rate). Table 9.3 shows the detection rates of these models generated by C4.5 when the false alarm rate is controlled at 1%. Models generated by other classifiers will achieve slightly different results [44].

We should point out that for certain attacks, especially the ones related to routing, it is not necessary to identify *every* abnormal event (or data point) in order to detect the attack because there may be *many* abnormal events caused by the attack. Therefore, we can use a post-processing procedure to count the number of detected abnormal events within each sliding time window, and

Table 9.3. Detection rates of the C4.5 models, with false alarm rate = 1%

Attacks	AODV	DSR
Blackhole	99%	87%
Selective dropping	95%	80%

Table 9.4. Features in the necessary conditions of the attacks

Blackhole	Packet Dropping
route add count	received ROUTE ERROR count
average route length	sent ROUTE ERROR count
received ROUTE REPLY count	sent HELLO count
	received HELLO count

conclude that an attack is present if the count is the *majority* or above a threshold. Using such a post-processing scheme, we can improve the detection rate and lower the false alarm rate. Another observation is that our detection models run at a frequency of the feature sample rate rather than continuously. They can potentially be the more efficient alternative than cryptography-based prevention scheme.

Efficiency. The models presented above use the full features set, which are clearly not energy efficient. We attempt a preliminary pre-pruning approach to reduce the number of features. The idea is to rank order the features based on their information gain [29], a measure on how much a feature contributes to classification. Our results show that by using just the top 15 features (versus the original 141 features), the detection models computed by C4.5 have very similar performance numbers as those shown in Table 9.3. We have just started the experiment in constructing simple detection modules for the cascaded detection scheme. We indeed find a number of necessary conditions of the attacks. The features in these conditions are shown in Table 9.4.

9.5 Conclusion

We have shown that the nature of MANET has instrinsic vulnerabilities which can not be removed. Evidently, various attacks that exploit these vulnerabilities have been devised and studied. New attacks will no doubt emerge in the future, especially when MANET becomes widely used. Defense against these attacks can be divided into two categories: attacks prevention and intrusion detection. While there are pros and cons in either category of techniques, they can work together to provide a better solution to address the security concerns. This is an important and still largely an open research area with many open questions and opportunities for technical advances.

References

[1] R. Agrawal, T. Imielinski, and A. Swami. Mining association rules between sets of items in large databases. In *Proceedings of the ACM SIGMOD Conference on Management of Data*, pages 207–216, 1993.

[2] D. Balfanz, D. K. Smetters, P. Stewart, and H. C. Wong. Talking to strangers: Authentication in ad-hoc wireless networks. In *Proceedings of the Network and Distributed System Security Symposium (NDSS)*, San Diego, CA, February 2002.

[3] S. Basagni, K. Herrin, D. Bruschi, and E. Rosti. Secure pebblenets. In *Proceedings of the 2001 ACM International Symposium on Mobile Ad Hoc Networking and Computing (MobiHoc 2001)*, Long Beach, CA, October 2001.

[4] J. Binkley and W. Trost. Authenticated ad hoc routing at the link layer for mobile systems. *Wireless Networks*, 7(2):139–145, 2001.

[5] R. Blom. An optimal class of symmetric key generation systems. In *Advances in Cryptology, EUROCRYPT'84, LNCS 209*, pages 335–338, 1984.

[6] L. Breslau, D. Estrin, K. Fall, S. Floyd, J. Heidemann, A. Helmy, P. Huang, S. McCanne, K. Varadhan, Y. Xu, and H. Yu. Advances in network simulation. *IEEE Computer*, 33(5):59–67, May 2000.

[7] S. Buchegger and J. L. Boudec. Nodes bearing grudges: Towards routing security, fairness, and robustness in mobile ad hoc networks. In *Proceedings of the Tenth Euromicro Workshop on Parallel, Distributed and Network-based Processing*, pages 403 – 410, Canary Islands, Spain, January 2002. IEEE Computer Society.

[8] S. Buchegger and J.-Y. L. Boudec. Performance analysis of the CONFIDANT protocol: Cooperation of nodes - fairness in dynamic ad-hoc networks. In *Proceedings of the IEEE/ACM Workshop on Mobile Ad Hoc Networking and Computing (MobiHoc)*, Lausanne, Switzerland, June 2002.

[9] H. Chan, A. Perrig, and D. Song. Random key predistribution schemes for sensor networks. In *Proceedings of the IEEE Symposium on Security and Privacy*, Berkeley, CA, May 2003.

[10] T. Clausen, P. Jacquet, A. Laouiti, P. Muhlethaler, and a. Qayyum et L. Viennot. Optimized link state routing protocol. In *Proceedings of IEEE International Multi-Topic Conference(INMIC)*, Pakistan, 2001.

[11] W. W. Cohen. Fast effective rule induction. In *Machine Learning: the 12th International Conference*, Lake Taho, CA, 1995. Morgan Kaufmann.

[12] J. R. Douceur. The sybil attack. In *Proceedings of the 1st International Workshop on Peer-to-Peer Systems (IPTPS'02)*, pages 251–260, March 2002. LNCS 2429.

[13] W. Du, J. Deng, Y. S. Han, and P. Varshney. A pairwise key pre-distribution scheme for wireless sensor networks. In *Proceedings of the 10th ACM Conference on Computer and Communications Security (CCS'03)*, October 2003.

[14] L. Eschenauer and V. D. Gligor. A key-management scheme for distributed sensor networks. In *Proceedings of the 9th ACM Conference on Computer and Communication Security*, Washington D.C., November 2002.

[15] Z.J. Haas and M. R. Pearlman. The zone routing protocol (ZRP) for ad hoc networks. Internet draft draft-ietf-manet-zone-zrp-04.txt, expired 2003, July 2000.

[16] C. Hsin and M. Liu. A distributed monitoring mechanism for wireless sensor networks. In *ACM Workshop on Wireless Security (WiSe)*, Atlanta, GA, September 2002.

[17] Y. Hu, D. Johnson, and A. Perrig. SEAD: secure efficient distance vector routing for mobile wireless ad hoc networks. *Ad Hoc Networks*, 1(1):175–192, July 2003.

[18] Y. Hu, A. Perrig, and D. Johnson. Ariadne: A secure on-demand routing protocol for ad hoc networks. In *Proceedings of ACM MOBICOM'02*, 2002.

[19] Y. Hu, A. Perrig, and D. Johnson. Packet leashes: A defense against wormhole attacks in wireless ad hoc networks. In *Proceedings of IEEE INFOCOM'03*, 2003.

[20] Y. Hu, A. Perrig, and D. Johnson. Rushing attacks and defense in wireless ad hoc network routing protocols. In *Proceedings of ACM MobiCom Workshop - WiSe'03*, 2003.

[21] Jean-Pierre Hubaux, L. Buttyan, and S. Capkun. The quest for security in mobile ad hoc networks. In *Proceedings of the 2001 ACM International Symposium on Mobile Ad Hoc Networking and Computing (MobiHoc 2001)*, Long Beach, CA, October 2001.

[22] S. Jacobs and M. S. Corson. MANET authentication architecture. Internet draftdraft-jacobs-imep-auth-arch-01.txt, expired 2000, February 1999.

[23] T. Joachims. *Making large-scale SVM learning practical*, chapter 11. MIT-Press, 1999.

[24] D. B. Johnson and D. A. Maltz. Dynamic source routing in ad hoc wireless networks. In Tomasz Imielinski and Hank Korth, editors, *Mobile Computing*, pages 153–181. Kluwer Academic Publishers, 1996.

[25] J. Kong, P. Zerfos, H. Luo, S. Lu, and L. Zhang. Providing robust and ubiquitous security support for mobile ad-hoc networks. In *Proceedings of the IEEE International Conference on Network Protocols*, Riverside, CA, November 2001.

[26] W. Lee. *A Data Mining Framework for Constructing Features and Models for Intrusion Detection Systems*. PhD thesis, Columbia University, June 1999.

[27] H. Mannila, H. Toivonen, and A. I. Verkamo. Discovering frequent episodes in sequences. In *Proceedings of the 1st International Conference on Knowledge Discovery in Databases and Data Mining*, Montreal, Canada, August 1995.

[28] S. Marti, T. J. Giuli, K. Lai, and M. Baker. Mitigating routing misbehaviour in mobile ad hoc networks. In *Proceedings of the Sixth Annual International Conference on Mobile Computing and Networking*, Boston, MA, August 2000.

[29] T. Mitchell. *Machine Learning*. McGraw-Hill, 1997.

[30] P. Papadimitratos and Z. J. Hass. Secure routing for mobile ad hoc networks. In *Proceedings of SCS Communication Networks and Distributed Systems Modeling and Simulation Conference (CNDS)*, San Antonio, TX, January 2002.

[31] C. Partridge, D. Cousins, A. W. Jackson, R. Krishman, T. Saxena, and W. T. Strayer. Using signal processing to analyze wireless data traffic. In *ACM Workshop on Wireless Security (WiSe)*, Atlanta, GA, September 2002.

[32] C. E. Perkins and P. Bhagwat. Highly dynamic destination-sequenced distance-vector routing (DSDV) for mobile computers. In *ACM SIGCOMM'94 Conference on Communications Architectures, Protocols and Applications*, pages 234–244, 1994.

[33] C. E. Perkins and E. M. Royer. Ad hoc on-demand distance vector routing. In *2nd IEEE Workshop on Mobile Computing Systems and Applications*, pages 90–100, New Orleans, LA, February 1999.

[34] A. Perrig, R. Canetti, J.D. Tygar, and D. Song. Spins: Security protocols for sensor networks. In *Proceedings of the Seventh Annual ACM International Conference on Mobile Computing and Networks (MobiCom 2001)*, Rome, Italy, July 2001.

[35] A. Perrig, R. Szewczyk, V. Wen, D. E. Culler, and J. D. Tygar. SPINS: security protocols for sensor networks. In *Mobile Computing and Networking*, pages 189–199, 2001.

[36] J. R. Quinlan. *C4.5: Programs for machine learning*. Morgan Kaufmann, San Mateo, CA, 1993.

[37] K. Sanzgiri, B. Dahill, B. N. Levine, C. Shields, and E. M. Belding-Royer. A secure routing protocol for ad hoc networks. In *Proceedings of ICNP'02*, 2002.

[38] M. Satyanarayanan, J. J. Kistler, L. B. Mummert, M. R. Ebling, P. Kumar, and Q. Lu. Experiences with disconnected operation in a mobile environment. In *Proceedings of USENIX Symposium on Mobile and Location Independent Computing*, pages 11–28, Cambridge, Massachusetts, August 1993.

[39] B. Schneier. *Secrets & Lies: Digital Security in a Networked World.* John Wiley & Sons, Inc., 2000.

[40] B. R. Smith, S. Murthy, and J.J. Garcia-Luna-Aceves. Securing distance-vector routing protocols. In *Proceedings of Internet Society Symposium on Network and Distributed System Security*, pages 85–92, San Diego, California, February 1997.

[41] F. Stajano and R. Anderson. The resurrecting duckling: Security issues for ad-hoc wireless networks. *Security Protocols. 7th International Workshop Proceedings, Lecture Notes in Computer Science*, pages 172–194, 1999.

[42] M. Zapata and N. Asokan. Securing ad hoc routing protocols. In *Proceedings of the ACM Workshop on Wireless Security (WiSe 2002)*, Atlanta, GA, September 2002.

[43] Y. Zhang and W. Lee. Intrusion detection in wireless ad-hoc networks. In *Proceedings of the 6th International Conference on Mobile Computing and Networking (MobiCom 2000)*, pages 275–283, Boston, Massachusetts, August 2000.

[44] Y. Zhang, W. Lee, and Y. Huang. Intrusion detection techniques for mobile wireless networks. *ACM Wireless Networks Journal*, 9(5):545–556, September 2003.

[45] L. Zhou and Z. J. Haas. Securing ah hoc networks. *IEEE Network*, 13(6):24–30, Nov/Dec 1999.

Index